PACKAGING DESIGN

Compiled and Edited by **Stanley Sacharow**

PBC International   New York, New York

PACKAGING DESIGN

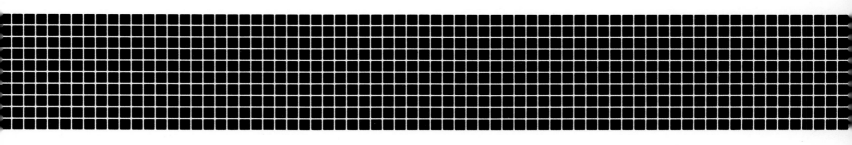

The Best of American Packaging and International Award Winning Designs

Distributors to the trade in the United States:
Robert Silver Associates
95 Madison Avenue
New York, NY 10016

Distributors to the trade in Canada:
General Publishing Co. Ltd.
30 Lesmill Road
Don Mills, Ontario, Canada M3B 2T6

Distributed in Continental Europe by:
Feffer and Simons, B.V.
170 Rijnkade
Weesp, Netherlands

Distributed throughout the rest of the world by:
Fleetbooks, S.A.
c/o Feffer and Simons, Inc.
100 Park Avenue
New York, NY 10017

Library of Congress Cataloging in Publication Data

Sacharow, Stanley.
   Packaging design.

   1. Packaging—Design and construction.   I. Title.
TS195.2.S2 1982   688.8   82-14487
ISBN 0-86636-000-X

Color separation, printing and binding
by **Toppan Printing Co., (H.K.) Ltd.**

Photo on title spread by Tom Tracy

10   9   8   7   6   5   4   3   2   1

To   Beverly Lynn
     Scott Hunter
     Brian Evan

Table of Contents

Foreword 8

Preface 10

Introduction 12

1.

Food Packages 14

Candy
Frozen Foods
Breads and Pastas
Meats
Dairy Products
Condiments
Baked Goods
Animal/Pet Foods

2.

Beverage Packages 74

Tea
Coffee

Soft Drinks
Beer
Wine
Liquor

3.

Cosmetics 106

Soaps
Body Care Products
Hair Care Products
Creams
Perfumes and Cologne
Makeup

4.

Drug/Health Related Products 136

Cold Relief Medications
Skin Care Products
Vitamins
Tissues

5.
Auto/Hardware Packages   158

Motor Oil
Car Care Products
Paints
Tools

6.
Tobacco Products   170

Cigarettes
Cigars

7.
Housewares   180

Paper Products
Detergents
Cleaners
Air Fresheners
Kitchen Equipment

8.
Potpourri   198

Toys
Office Supplies
Cameras/Camera Equipment
Stereo Equipment
Apparel
Sporting Goods
Shopping Bags/Boxes

9.
Award Winning Packaging   222

Aesthetic Award Winners
Industry Award Winners
International Award Winners

Appendix   252

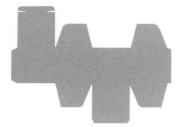 # Foreword

With all of the splendid "tabletop" books available, it's interesting to note that so few have devoted their excellent design, photography, and printing to a subject that is an integral part of the American existence—packaged products. In a country where material purchases have been the essence of the good life (as well as survival) where large sums of money are spent to obtain the best marketing and design for the greatest return— resulting in the most exciting and best planned packaging and advertising anywhere—it is surprising how casually the multitudinous products of our daily life are perceived and accepted.

Many packages are the epitome of fine design. That they achieve their marketing function—to be displayed and bought—is a foregone conclusion, or they would not remain on the store shelves. To view each objectively as a relationship of color, space, graphics, proportion, materials, reproduction, is mainly a function left solely to designers and their clients. Therefore this book has brought together over 600 examples of outstanding packaging so that everyone in the industry—marketers, designers, manufacturers, production people; plus students and interested neophytes—can

see what is possible. It is a showcase for the efforts of my colleagues and professional designers. The selective fruits of their talent, taste, and attention to detail are shown here for all to study, learn from and—above all enjoy. Just as each package is the result of the concerted work of a team of experts and knowledgeable practitioners, so it.was with the production of this book.

MAY BENDER, *President*
May Bender Industrial Design, Inc. *and*
Past President
of Package Designer's Council

"Design is a function of merchandising, distribution and other factors, tempered by the restrictions of manufacturing."

ROY PARCELS
Dixon and Parcels Associates, Inc.

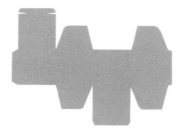

The Evolution of Packaging

"The principles discoverable in the works of the past belong to us; not so the results. It is taking the ends for the means . . . . No improvement can take place in the Art of the present generation until all classes, Artists, Manufacturers and the Public, are better educated in Art, and the existence of general principles more fully recognised."

Owen Jones
The Grammar of Ornament (1854)

As long ago as 1940, in his work on art education, *Academies of Art Past and Present,* Sir Nikolaus Pevsner observed that "a history of art could be conceived not so much in terms of changing styles as of changing relations between the artist and the world surrounding him."

A few years earlier, the visionary designer Norman Bel Geddes had seen the opportunities offered by industry. In his book *Horizons* he wrote, "The few artists who have devoted themselves to industrial design have done so with condescension, regarding it as a surrender to Mammon, a mere source of income to enable them to obtain time for creative work. On the other hand, I was drawn to industry by the great opportunities it offered creatively."

Whether a package for food or an industrial 55-gallon drum, there is nothing inevitable about the appearance of manufactured objects. They didn't *have* to look this way or that.

Design is what occurs when art meets industry, when people begin to make decisions about what mass-produced products should look like. It was because Britain faced the artistic and social consequences of the Industrial Revolution before any other country that design developed there first.

In the eighteenth century there was a widely felt impulse to rationalize thought, to inquire into the nature of things and to quantify results. This sort of investigation led the painter William Hogarth to produce his book *The Analysis of Beauty,* in which he attempted to set out the laws which govern our responses to art.

At the same time, the potter Josiah Wedgwood was coming to terms with the reality of the early Industrial Revolution. The new manufacturing and commercial processes had separated inventing from making and making from selling. A century before Marx, Wedgwood had already perceived the inevitability of the division of labor, and found that this was not always to the benefit of either art or industry. By employing practicing artists like Stubbs and Flaxman, he introduced the designer to industry. They sent the perfumed breath of art, often inspired by the past, into the sulphurous, volcanic heart of industry. Wedgwood inspired many imitators, but besides calling into existence this new class of being, the designer, he made distinguished contributions to mass-production by developing new techniques and taking prodigious advantage of the commercial benefits offered by the division of labor.

Wedgwood was, perhaps, more

inspired by the mercantile than the "moral" aspects of design. The equivocation he must have felt is indicated by his distinction between "useful" and "beautiful" ware, but both mercantile and moral opportunities offered by the integration of art into industry were simultaneously sensed at an official level: in its ambitious program to unite commerce and the arts, the Society of Arts summarized the material and metaphysical aims of the age. In the Society of Arts, Britain matched its international industrial lead with an institution invented to administer the Industrial Revolution.

However, it is a sad irony that this institution, created in the Age of Enlightenment, had effects more cerebral than practical. Eighty years after the founding of the Society of Arts, a Parliamentary Select Committee was established "to inquire into the best means of extending a knowledge of the Arts and of the Principles of Design among the People (especially the Manufacturing Population)." The Committee heard evidence from distinguished foreigners, and when its report was eventually published in 1836 the conclusion was that standards of design were higher abroad than at home and that the Committee must intensify its efforts "to infuse, even remotely, into an industrious and enterprising people a love of art."

This motivation was the beginning of the government sponsored Schools of Design in Ornamental Art. This was also the first age of the museum, when it was believed that gathering together artifacts of high quality from across the ages would necessarily inspire the same quality in current production. In 1837 the old premises of the Royal Academy in Somerset House were made available for the School of Design (a substantial symbol of fine art moving over for design in the new machine age) and soon seventeen other local Schools of Design were set up in the provinces.

A further lesson in taste was presented to the public by Prince Albert and his circle which resulted, in 1851, in the creation of the Great Exhibition of the Industry of all Nations. Conceived as ameliorative instruction, the Great Exhibition proved to be a real shock. It was seen that everywhere industry was out of control. Mass-produced objects were scarred by vulgar and inappropriate ornament; prophetically, the German architect Gottfried Semper noted that only functional objects—such as carriages and guns—seemed to be well designed.

The result of the Great Exhibition was that sensitive individuals close to Prince Albert were prompted to analyze esthetic principles and to generalize from them. The most remarkable instance of this was a book, *The Grammar of Ornament*, published by a Welsh architect, Owen Jones,* in 1854. Jones, who also believed that every town should have an art gallery, provides us with the germ of an idea that there is an appropriate finish for every material and an appropriate form for every machine.

The other consequence of the Great Exhibition, whose idea was continued in another exhibition of 1862, was the creation of what ultimately became the Victoria and Albert Museum, which housed the bulk of the surplus exhibits. Henry Cole combined the School of Design into the new Museum of Practical Art—changing its name to the South Kensington Museum in 1859. It was renamed the Victoria and Albert in 1899.

Henry Cole's Museum was remarkable because he was prepared to display badly designed objects, thus inverting the ideas of the classical academy. These were not displayed so that they may be imitated, but so that study of them might excite some higher ambition. This idea of a useful, practical museum spread across Europe—to Vienna, Karlsrühe, Berlin, Cologne, Munich, Hamburg and Oslo.

By the mid-1920s, all the ingredients necessary for the proper development of a package were available. Needed still

was some added marketing refinement in the area of consumerism. That was soon to come with the work done at the Harvard Business School in the 1940s and the introduction of the concept of "the marketing mix." When supermarkets arrived on the scene and package design became a vital selling force, the stage was set for what Daniel J. Boorstin called "one of the most manifold and least noticed revolutions in the common experience."

In the United States, starting about 1900, production-minded executives took hold of this new design force and saw the opportunity for using both product improvement and product appearance as two forceful selling points. They sensed the value of these two factors as selling tools for increasing sales volume through advertising and sales promotion.

There soon developed concepts of "brand loyalty," and the old world of "hallmarks" rapidly became transformed into the new world of trademarks and brand names. The first federal trademark law was passed in 1870, but a decade later was declared unconstitutional because it purported to restrict commerce within the states. A new law passed in 1881 became the basis for the registration and protection of all trademarks used in interstate and foreign commerce. Brand names had acquired a new power and a new meaning.

This was the beginning of an attempt to establish rules for what manufactured objects should look like. It was a practical age. It was a moment when industry had demonstrated its power and now looked for inspiration.

STANLEY SACHAROW
E. BRUNSWICK, N.J.

---

*Owen Jones designed various biscuit tins for Huntley and Palmers; and was responsible for their famous "Caslet" tin.

Introduction

Package Design in the 1980s

Trends in package design often reflect worldwide social and economic conditions, and the designs of the early and middle 1980s are certainly no exception. With a world recession raging and a discernible trend by consumers toward economy, packaging—particularly food packaging—has been characterized by a plain, simple and bold look. Copy has been restyled to unsnarl the jumble of words that had been added during the 1970s to satisfy government regulations. More direct cues as to the packaged contents—often in the form of striking photos of the product itself (particularly in food packaging)—have been moved to the front panels from the backs and sides. Simple graphics aimed at provoking shelf-impact legibility is the norm in the 1980s. (Recent studies by Walter Margulies of Lippincott and Margulies have shown that one out of every six shoppers who needs eyeglasses does not wear them while shopping.) But the economy more than any single factor is altering package design. Writes James Shannan in the *Wall Street Journal* (August 8, 1981), "Inflation has pushed people to wring the last ounce of energy from their marketing tools."

The economy has been a major consideration for generics ("no brand" products) and house brands capturing about 5 percent of the $200 billion American grocery market. It is predicted that generics could capture as much as 25 percent of the market in the 1980s. The relatively sparse generic labels, used to convey an impression of economy to the consumer, vary very little among different generic brands. Although most generics

use simple black and white labels, there are several types on the market with more elaborate labels, and one Canadian supermarket features generics with yellow and black labels. The trend toward one generic looking "better" than another appears to have already started. How and when it will mature is anybody's guess.

The consumer in the 1980s is a health-conscious individual, concerned about the "lightness" of a product. Because of the national passion for slimness in the U.S., product manufacturers have increased the development and introduction of "light" foods and beverages. Most of these products feature lots of white background with sky blue or gold highlights. The use of variety lettering also serves to connote both quality and lightness.

Among the more successful products introduced in 1981–82 are sugarless cereals, wheat-soy spaghettis and low-calorie frozen entrees; all are packaged in neutral colors with simple, uncluttered graphics. Neutral and earth tones project both health and economy to consumers, and these are frequently used by manufacturers with considerable success.

Energy, a critical factor in the mind of the consumer in the 1980s, played an important role in a new package concept introduced in 1981. Unfortunately the retort pouch (a flexible, 3-ply lamination pouch, that is shelf-stable and meant to replace the conventional tinplate can), though a definite energy-saver, failed to capture the American consumer's imagination. In its place has arrived another concept, already used in Europe, that promises to change the buying habits of the American shopper. Aseptic packages (packages made of sterile material and combined with the product in a sterile environment) of juice, soft drinks and milk were introduced to the American scene after being extensively used in both Europe and Asia. These square Brik-Paks and Ready-Paks offer outstanding opportunities to the designer. Not only must the package designer attempt to promote a "stand-out" quality through graphics, he or she also has the opportunity to introduce a degree of "Europeanism" into a rather staid American product mix. One juice line introduced by a major food corporation in 1981 does exhibit superb graphics, and it is anticipated that further excellent designs in this area will continue to be seen.

The recession has forced many established firms to seriously consider package redesigns for their current products. After all, it was during, and because of the Great Depression of the 1930s, that the then fledgling package design industry grew to maturity and became a vital force in American business. The design of packaging is now—and increasingly will be—one of the really critical ways in which the manufacturer can communicate his ideas to the customer. There is a definite and important trend toward the upgrading of existing design themes. Many well-established and conservative product designs have had to modernize in order to keep up with current trends, consumer demands and the ever-increasing competition within the industry.

A popular trend in package redesign is the use of photographs in place of artwork. One large manufacturer of animal foods has successfully capitalized on the fact that pet-owners attribute human traits to their pets by replacing drawings of these animals with photographs of dogs and cats with people.

In the cosmetics area, the bolder sexual designs of the late 1970s are being refined and given new direction. Perhaps because of the current trend (at least among certain outspoken groups) toward sexual conservatism, cosmetics packaging has begun to take on a newer, more discreet look. Sports are featured, especially on men's toiletries. Women's personal hygiene aids and cosmetics feature designs that are less blatant and more softly, romantically sensual. There is also a growing tendency to intensify the impact of cosmetics packages through vivid colors and bold shapes in geometric or sculptural configurations.

Finally, it has become apparent in the 1980s that the products sold in supermarkets more or less depend on the value of packaging for their success. The supermarket has become a visual jungle, and good design is needed to help the consumer get through its aisles with as little confusion as possible. What is still less widely appreciated (although this is rapidly changing) is the power of good package design to speed the distribution and sale of products through other outlets. Still needed are good package designs for hardware, tools, clothing and soft goods—designs that tease the imagination (and open the pocketbooks) of the consumer.

Pressed by tough economic times and steadily increasing competition, the package designs of the 1980s are a dynamic mixture. They are being asked to stimulate the sales of stagnant brands and often to perform the role of actually advertising the product. Worldwide, package design is certainly living up to its goals.

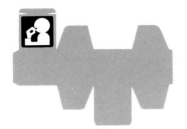

1

Food Packages

"American food processors spent more than $23 billion last year to make their products look appetizing in their containers, according to the United States Department of Agriculture. That represents about 8 percent of the retail cost of food."

The New York Times
April 14, 1982

Packages and their products are often developed for specific markets that are determined by definite demographic trends. Several of the trends evident in food products and packaging in the 1981–82 period are discussed below.

Maturity Market

According to *U.S. News and World Report*, by the year 2000 the 35–49 age group will have increased by 76 percent of present figures, with a 53 percent increase in the population of those over the age of 50.

The stress on the "natural" qualities of cereals, breads and cookies has been, at least in part, a direct result of manufacturers' interest in capturing two growing markets: the mature (45–64) and older (over 64) age groups. Many recently introduced international packages have featured the wheat or bran content of the product on the container. This is a direct attempt to persuade purchasers that what is wholesome is nutritionally correct. In the United States, many food packages are displaying grains to impart a feeling of "goodness." In the United Kingdom, the packages designed by the Tyrell Company for Wheat Heart, Bran Brek and Day Brek use shades of brown and gold and illustrations of attractive sunlit scenes from the English countryside to emphasize "how good the product is for you."

Another popular trend is the increasing appearance of "light" foods—foods with fewer calories that are lower in sugar and fats. For such food lines, white and off-white seem to be the colors of preference. The low-sugar canned fruit lines of two major American food processors use white predominantly. Another large company has brought out a new food line containing approximately 30 items with a rather spartan package design on a deceptively rich-looking cream background. The idea is to make the food look simple and appetizingly healthful, rather than bland and puritanically dietetic.

With improved health care, a dwindling birthrate, and the aging of the Post-World War II boom babies, the maturity market will continue to increase. Some recent products developed for this age group include high fiber cereals, low cholesterol cheeses, low calorie frozen entrees and chewing gum that will not stick to dentures. A major food corporation expects that sales of prune juice will rise as people grow older (possibly causing several dramatic new innovations in juice-package design).

Families

In 1950, the average size of the American family was 3.54 individuals. By 1970, it was little changed—3.57. Then, in the

1970s, family size took a sharp drop to 3.27, partly because of a steep increase in one-parent families and partly because of lower birth rates. By 1970, 12.3 percent of all American families were headed by a single parent. By 1980 the percentage had increased dramatically to 19.1.

Childless households are also on the rise. Young, two-income couples tend to be more affluent, more mobile, more oriented toward immediate pleasures, and more interested in leisure-time activities and fashions than the more traditional family households. There is now a growing market among this consumer group for fancy foods, convenience foods, single-serving entrees and foods packaged in attractive containers specifically designed to capture the consumer's imagination.

In the United Kingdom, there is also a trend toward the increased consumption of prepared convenience foods. Many excellent design examples can be found on frozen food, snack and soup packages.

Mobility and Regionalism
At the turn of the century, 78.8 percent of the American people resided in the states in which they were born. That percentage declined slightly but steadily until 1970, when it reached 68 percent. During the 1970s the drop was considerably more precipitous than in the past, and by 1980 the figure was down to 63.8 percent. Officials believe this trend is largely because many people moved from the Northeast and Middle West in that period for better job opportunities (and a more pleasant climate) in the South and West. Also, many people tended to retire to areas far from their homes.

With more people moving to warmer climates, this will mean an increase in the use of warmer colors and designs associated with those regions. It will also increase the demand for foods that can be prepared outdoors, such as barbecue sauce and frankfurters. While bright, warm colors are preferred by residents of Florida and the Southwest, grays and cooler colors still predominate in New England and the Midwest.

Regionalism is also an important factor in international package design. In Scotland, people spend proportionately more of their disposable incomes on shoes and tobacco than do those in Wales, where the largest expenditures are made on services and household durables. As Tom Sutton of J. Walter Thompson has observed, "No product, no brand, can be everything to all men."

The Search for Novelty
As the American diet becomes more and more catholic, the search becomes ever wider for new and different foods, or for new combinations of known comestibles. It is significant to note that June 1982 was the biggest month for new food product introduction since Dancer Fitzgerald Sample's *New Product News* began monitoring these trends in 1964. Over 152 new food and drug products were introduced in that one month, up from 144 products a year earlier. The June tally pushed new introductions for the first half of 1982 to a record 684 products.

Among the June introductions were Tofu Lasagna, a blending of cuisines of East and West that substitutes a soybean product for the normal cheese. Another newcomer is Tater Nuts, a potato chip wrapped around a peanut in a combination of two highly popular snack foods. With new products such as these are sure to come a wide variety of innovative and dramatic packages. If the trend continues, 1982 will certainly be a banner year for food packaging.

Nutritional Labeling
There is a discernible trend among manufacturers of food products toward the inclusion of nutritional listings, especially sodium content, on their packages. This is obviously a response to increased consumer interest in the foods they eat, particularly foods of the convenience and "ready to eat" variety. On the other hand, this data, coupled with federal requirements for full-ingredient disclosure, has the effect of "busying" the label and thereby creating both esthetic and logistical problems for the package designer. There is definite concern among people in the packaging industry that too much information is on the label. With the arrival of the Reagan administration, there is some reason for hope among members of the industry that further governmental legislation in this area will be at least partially curtailed.

Like all new products, new food products must, if they are to be successful, inspire confidence in the consumer. The precise way the product indicates this to the consumer may vary— from a concentration on ingredients (natural foods) through stress on ease of preparation (convenience foods) to emphasis on lifestyle (prepared nonrefrigerated entrees). But in most cases the package determines whether the consumer will come away from the supermarket convinced or unimpressed. As David Ogilvy once pointed out, "The greater the similarity between products, the less part reason plays in brand selection."

1. Product: Confection
   Design Firm: Oubaku
   Client: Kagiya Masaaki Inc.

2. Product: Spring Gum
   Art Director: Gaylord Adams/Donald Flock
   Designer: Gaylord Adams/Donald Flock
   Design Firm: Gaylord Adams Associates, Inc.
   Client: Warner-Lambert

3. Product: Chocolate Card
   Art Director: Jack Schecterson
   Designer: Jack Schecterson
   Design Firm: Jack Schecterson Associates Inc.
   Client: Astor Chocolate Corp.

4. Product: Petit Francais Candy
   Art Director: Takeo Yao
   Design Firm: Yao Design Institute Inc.
   Client: Francais Confectionery Co. Ltd.

5. **Product:** Shishi Confection
   Art Director: Kuniie Misawa
   Designer: Takuji Takahashi
   Client: Eitaro Sohonpo Inc.

6. **Product:** Old Country Candy Box for Busch Gardens at Williamsburgh, VA
   Graphic Designer: Robert Bowers; Robert A. Bowers Graphics
   Structural Designer: H. Winstead Jones; Pohlig Brothers, Inc.
   Client: Busch Entertainment Corp.

7. **Product:** Fruit Cocktail Candy
   Art Director: Takeo Yao
   Design Firm: Yao Design Institute Inc.
   Client: Francais Confectionery Co. Ltd.

8. **Product:** Carob Bars
   Structural Design: Fred Mark
   Design Firm: Lebanon Packaging Corporation
   Client: Burry Health Foods Supply, Inc.

9. **Product:** Tea Candy
   Art Director: Takeo Yao
   Design Firm: Yao Design Institute Inc.'
   Client: Eitaro Sohonpo Inc.

10. **Product:** Confection
    Design Firm: Seijaku
    Client: Oharameya Inc.

11. **Product:** Confection
    Design Firm: Tama Azuki
    Client: Kinukake Confectionery

12. **Product:** Gubor Privat Chocolates
    Design Firm: Graphia Hans Gundlach
    Client: Gubor Schokoladen Fabrik

13-15. **Product:** Au Chocolat Boxes and Shopping Bag
Designer: Robert P. Gersin
Design Firm: Robert P. Gersin Associates Inc.
Client: Bloomingdale's

16. **Product:** Moyet Candy
    Art Director: Shigeo Okamoto
    Designer: Shigeo Okamoto
    Client: Francais Confectionery Inc.

17. **Product:** Confection
    Designers: Hirome & Gosho no Hana
    Client: Matsumaeya Inc.

18. **Product:** Odile Candy
    Art Director: Valentine F. Morozoff
    Designer: Kosaku Murata
    Client: Cosmopolitan Confectionery

19. **Product:** Whitman's Sampler
    Art Director: Ed Morrill
    Designer: J.D. Grinnell
    Design Firm: Werbin & Morrill Inc.
    Client: Whitman Division of Pet, Inc.

20. **Product:** Fruit Bag Gift Set
    Art Director: Takeo Yao
    Design Firm: Yao Design Institute Inc.
    Client: Maruemu A. Coop

21. **Product:** Carefree Gum
    Art Director: Gaylord Adams/Donald Flock
    Designer: Gaylord Adams/Donald Flock
    Design Firm: Gaylord Adams and Associates Inc.
    Client: Beechnut Lifesavers, Inc.

22. **Product:** Brandy Cake
    Art Director: Takeo Yao
    Design Firm: Yao Design Institute Inc.
    Client: Francais Confectionery Inc.

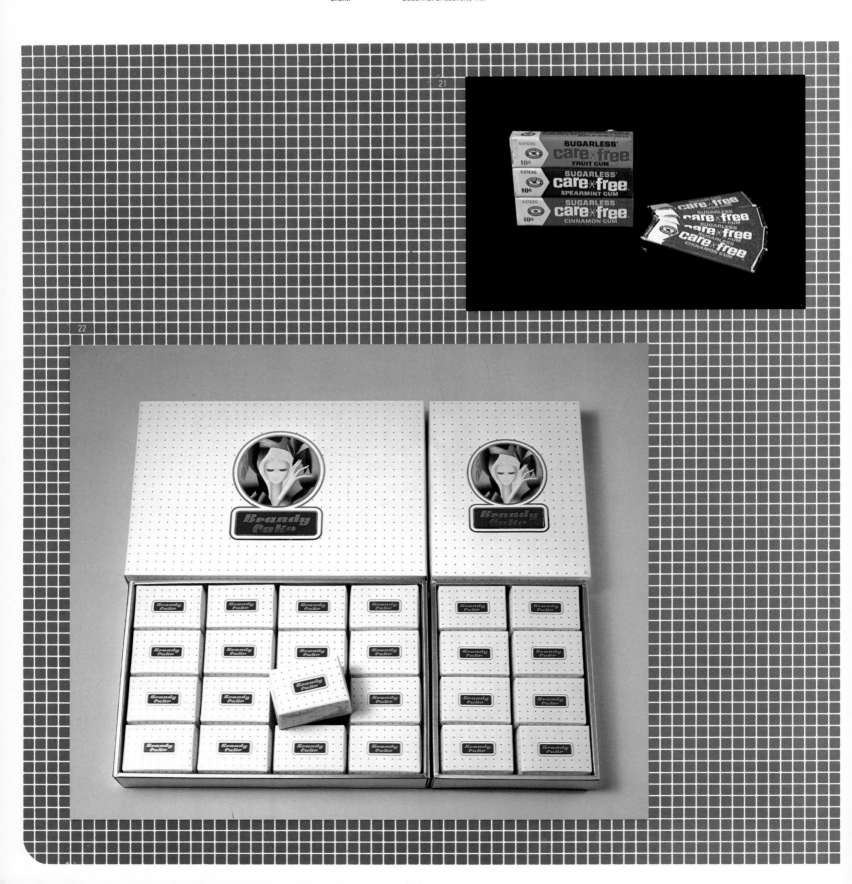

23. **Product:** Chocolate Dessert Shells
    Art Director: Jack Schecterson
    Designer: Jack Schecterson
    Design Firm: Jack Schecterson Design Associates Inc.
    Client: Astor Chocolate Corp.

24. **Product:** Cella Cherries
    Art Director: Gaylord Adams/Donald Flock
    Designer: Gaylord Adams/Donald Flock
    Design Firm: Gaylord Adams & Associates Inc.
    Client: Cella's

25. **Product:** Chocolate Chess Set
    Art Director: Jack Schecterson
    Designer: Jack Schecterson
    Design Firm: Jack Schecterson Associates Inc.
    Client: Astor Chocolate Corp.

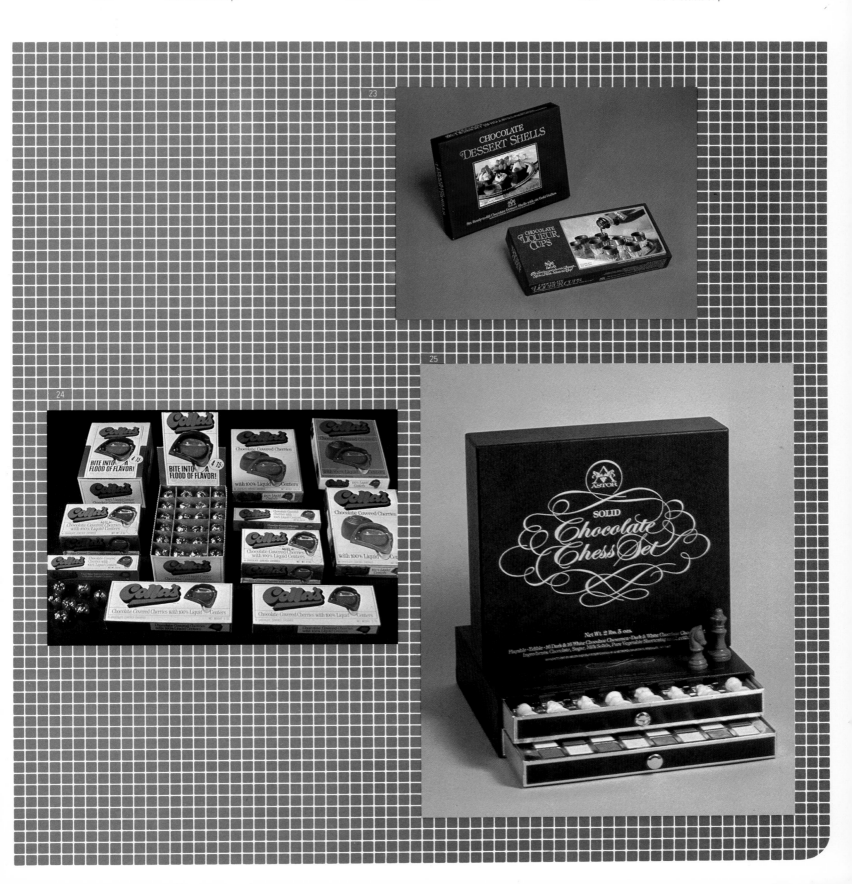

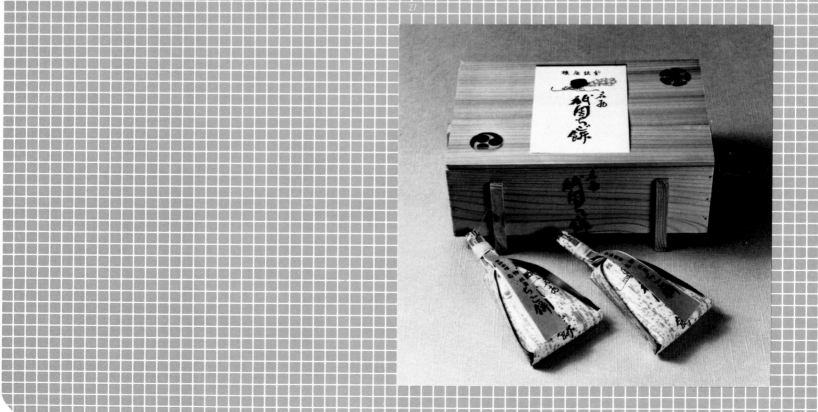

26. **Product:**   Christmas Candy Box Ornament
    Art Director: Timothy J. Musios
    Designer: Curtis L. Iverson/Charles W. Duncan
    Client: Avon Products, Inc.

27. **Product:**   Confection
    Designer: Gian Chigo Mochi
    Client: Sanjo Wakasaya, Inc.

28. **Product:**   Cresta Red Chili Burritos
    Designer: Doris Ray
    Design Firm: Harte Yamashita & Forest
    Client: Dob Foods

29. **Product:**   Hors D'Oeuvre Tray
    Design Firm: Dixon & Parcels Associates, Inc.
    Client: Bernan Foods

30

31

32

30. **Product:** Chun King Frozen Food Line
    Creative Director: John S. Blyth
    Designer: Penny Johnson
    Design Firm: Peterson & Blyth Associates Inc.
    Client: R.J. Reynolds Industries, Inc.

31. **Product:** Mrs. Smith's Quiche
    Structural Design: Jim Capo
    Design Firm: International Paper Company
    Client: Mrs. Smith's Frozen Foods Company

32. **Product:** Patio Frozen Mexican Entrees
    Design Firm: Dixon & Parcels Associates Inc.
    Client: R.J. Reynolds Industries, Inc.

33. **Product:** Michel Guérard Comptoir Gourmand logo, shopping bags, wrappers and labels
    Designer: Andrew Herz
    Design Firm: Robert P. Gersin Associates Inc.
    Client: Bloomingdale's

34. **Product:** Mrs. Paul's Pierogies
    Design Firm: Dixon & Parcels Associates Inc.
    Client: Mrs. Paul's Kitchens, Inc.

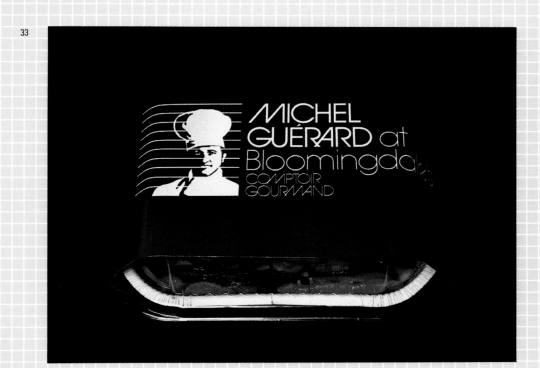

33

34

35. Product: Gorton's Frozen Seafood
Creative Director: John S. Blyth
Designer: Penny Johnson
Design Firm: Peterson & Blyth Associates Inc.
Client: Gorton's of Gloucester

36. Product: LaChoy Frozen Chinese Food Line
Design Firm: Dixon & Parcels Associates Inc.
Client: La Choy Food Products

37. Product: Mrs. Weinberg's Fried Rice
Design Firm: Martin Jaffe Design Inc.
Client: Mrs. Weinberg's Food Products Corp.

38. Product: Celeste Chicago Style Pizza
Account Manager: Owen W. Coleman
Designer: Owen W. Coleman/Abraham Segal/John Rutig
Design Firm: Coleman, LiPuma & Maslow, Inc.
Client: The Quaker Oats Company

39-40. Product: Marcella Hazan's Italian Kitchen label, shopping bag, boxes, containers
Designer: Candace Cain
Design Firm: Robert P. Gersin Associates Inc.
Client: Bloomingdale's

38

39

40

41

42

43

41. Product: Totino's Heat and Eat Pizza
    Art Director: Richard Deardorff
    Designer: Jan Rosamond
    Design Firm: Overlock Howe Consulting Group, Inc.
    Client: Pillsbury Co.

42. Product: Birdseye Frozen Vegetables Polybags
    Design Firm: Charles Biondo Design Associates, Inc.
    Client: General Foods Corporation

43. Product: Birdseye Blue Ribbon Combinations Line
    Account Manager: Sal V. LiPuma
    Designers: Sal V. LiPuma, Ward M. Hooper
    Client: General Foods Corporation

44. Product: Gorton's New England Clam Chowder
    Creative Director: John S. Blyth
    Designer: Penny Johnson
    Design Firm: Peterson & Blyth Associates Inc.
    Client: Gorton's of Gloucester

45. Product: Skillet Magic
    Art/Creative
    Director: John DiGianni
    Design Firm: Gianninoto Associates, Inc.
    Client: McCormick & Co., Inc.

46. Product: Cup-a-Soup
Art/Creative
Director: John DiGianni
Design Firm: Gianninoto Associates, Inc.
Client: Thomas J. Lipton, Inc.

47. Product: Cup Noodle
Art Director: Takeshi Otaka
Designer: Takeshi Otaka
Client: Nisshin Food Co., Ltd.

48-49. Product: O My Goodness Soun
Designer: Robert P. Gersin/Georgina Leaf
Design Firm: Robert P. Gersin Associates Inc.
Client: Myojo Foods of America

50. Product: Bell's Meatloaf Mix
Art Director: Ed Morrill
Designer: Ed Morrill
Design Firm: Werbin & Morrill Inc.
Client: Brady Enterprises

47

48

49

50

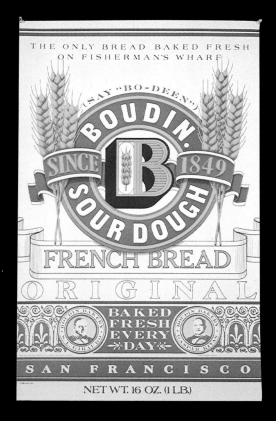

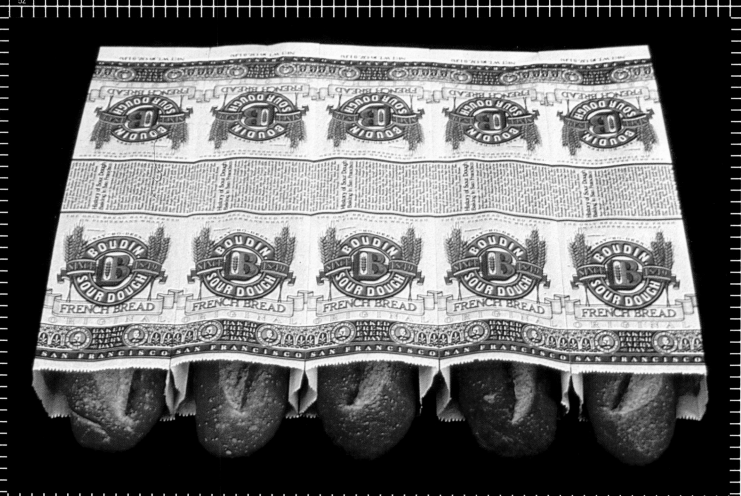

51-52. Product: Boudin's Sour Dough Bread
Art Director: Primo Angeli
Designer: Primo Angeli
Design Firm: Primo Angeli Graphics
Client: Boudin Bakeries

53. Product: Born Brown Bread Series
Art Director: Hiroshi Abe
Designer: Hideo Hama
Client: Pasco Inc.

54. Product: Country Hearth Bread
Design Firm: Dixon & Parcels Associates Inc.
Client: Butter Krust Bakeries

55. Product: Home Pride Bread
Design Firm: Dixon & Parcels Associates Inc.
Client: ITT Continental Baking Co.

56. **Product:** Goshiki no Ito (noodles)
    Art Director: Takeshi Tachi
    Designer: Hiroyuki Veno/Shizue Hayashi
    Client: Morikama Inc.

57. **Product:** 4C Bread Crumbs
    Design Firm: Dixon & Parcels Associates Inc.
    Client: 4C Foods Corp.

58. **Product:** Fruit & Fibre Cereal
    Design Firm: Murtha, DeSola, Finsilver, Fiore Inc.
    Client: General Foods

59. **Product:** Leonardo Pasta Shells
    Art Director: Ray Peterson
    Designer: Marvin Steck
    Design Firm: Container Corporation of America
    Client: Noodles by Leonardo

60. **Product:** All Ready Pie Crusts
    Creative Director: John S. Blyth
    Designer: Penny Johnson
    Design Firm: Peterson & Blyth Associates Inc.
    Client: Pillsbury Co.

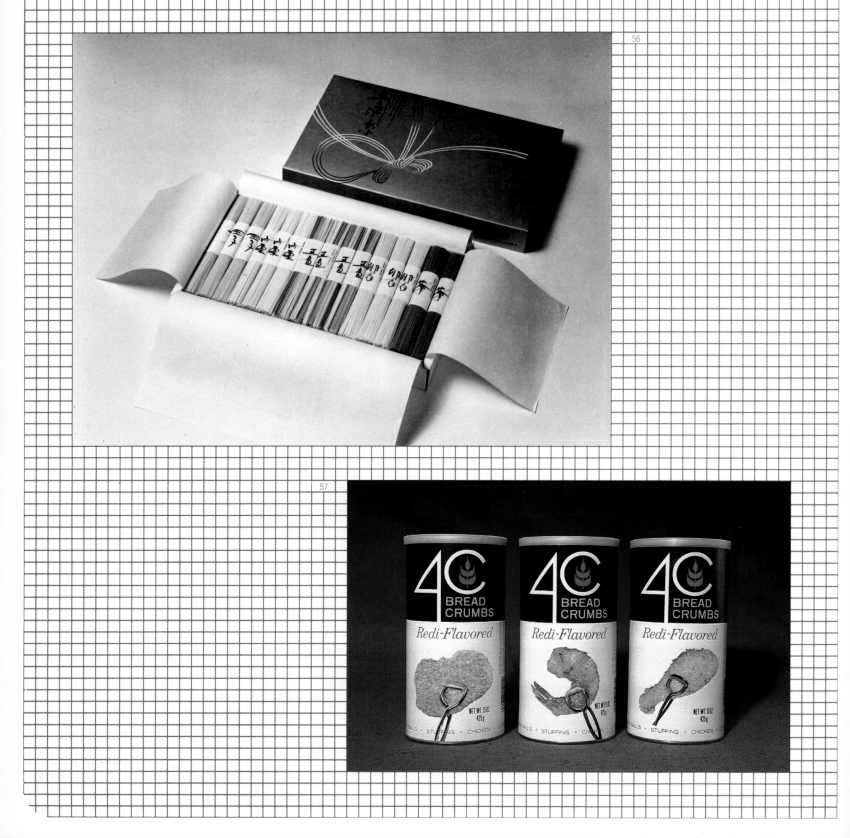

58

59

60

61. **Product:** Dak Ham
Art Director: John DiGianni
Creative Director: John DiGianni
Design Firm: Gianninoto Associates Inc.
Client: Dak Foods Inc.

62. **Product:** Tulip Ham
Design Firm: Schur-Verpackungen
Client: Tulip Slagterierne

63. **Product:** Country Cookhouse Corned Beef
Art Director: Donald Flock
Designer: Donald Flock
Design Firm: Gaylord Adams & Associates Inc.
Client: Caemint Foods

64. Product: Crisp 'n Lean
    Art Director: May Bender
    Designer: May Bender
    Design Firm: May Bender Industrial Design
    Client: Horace W. Longacre, Inc.

65. Product: Louis Rich Turkey products
    Art Director: Richard Deardorff
    Designer: Richard Deardorff
    Design Firm: Overlock Howe Consulting Group, Inc.
    Client: Oscar Mayer Co.

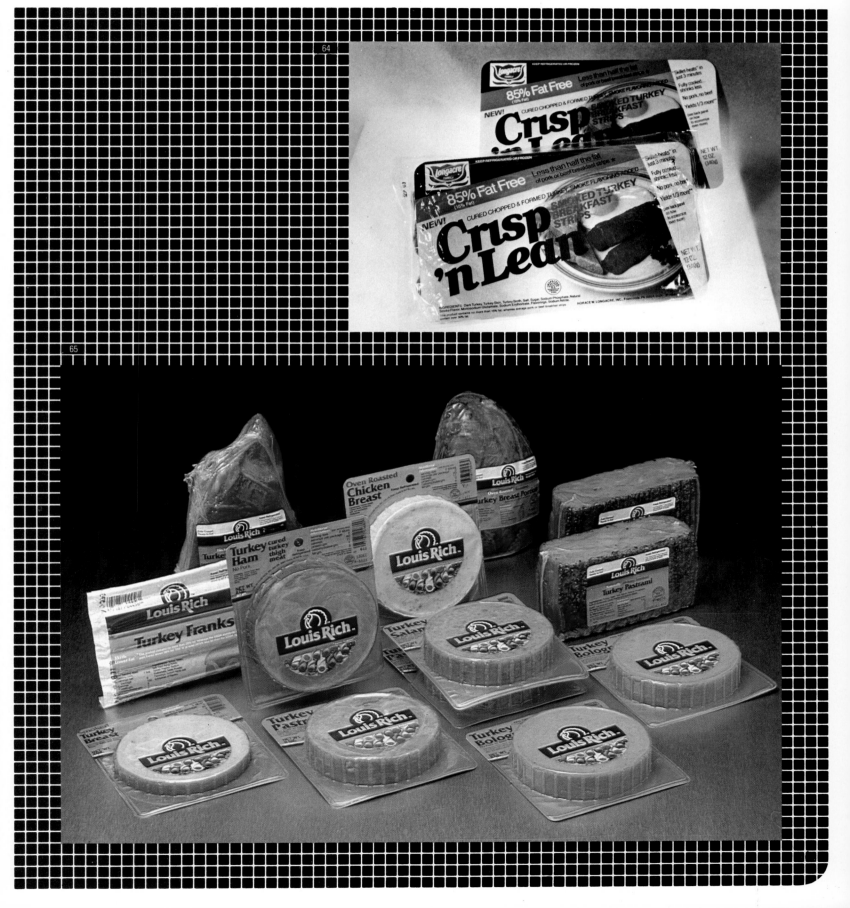

66. **Product:** Dorman's Cheese
    Creative Director: Clive Chajet
    Design Firm: Chajet Design Group, Inc.
    Client: Dorman Cheese Co.

67. **Product:** Snow Brand Royal Gift Set
    Art Director: Takeo Yao
    Design Firm: Yao Design Institute Inc.
    Client: Snow Brand

68. **Product:** Friendship Dairy Line
    Design Firm: Martin Jaffe Design Inc.
    Client: Friendship Dairies, Inc.

69. **Product:** Friendship Yogurt Line
    Design Firm: Martin Jaffe Design Inc.
    Client: Friendship Dairies, Inc.

70. **Product:** Apple & Raisin Butter
    Art Director: Takeo Yao
    Design Firm: Yao Design Institute Inc.
    Client: Snow Brand

68

69

70

71

73

72

71. Product: Crosse & Blackwell Gourmet Product Line
    Account Manager: Owen W. Coleman
    Designer: Owen W. Coleman/Abraham Segal
    Design Firm: Coleman, LiPuma & Maslow, Inc.
    Client: Libby, McNeill, Libby, Inc.

72. Product: Pet Ice Cream
    Design Firm: Charles Biondo Design Associates, Inc.
    Client: Pet Dairy Division

73. Product: Old Philadelphia Ice Cream
    Art Director: Paul Gee
    Creative Director: Irv Koons
    Design Firm: Irv Koons Associates, Inc.
    Client: Abbott's Dairies

74. Product: Good Humor Ice Cream
    Art Director: John DiGianni
    Creative Director: John DiGianni
    Design Firm: Gianninoto Associates Inc.
    Client: Thomas J. Lipton, Inc.

75. Product: Pet Specialty Ice Cream
    Design Firm: Charles Biondo Design Associates Inc.
    Client: Pet Dairy Division

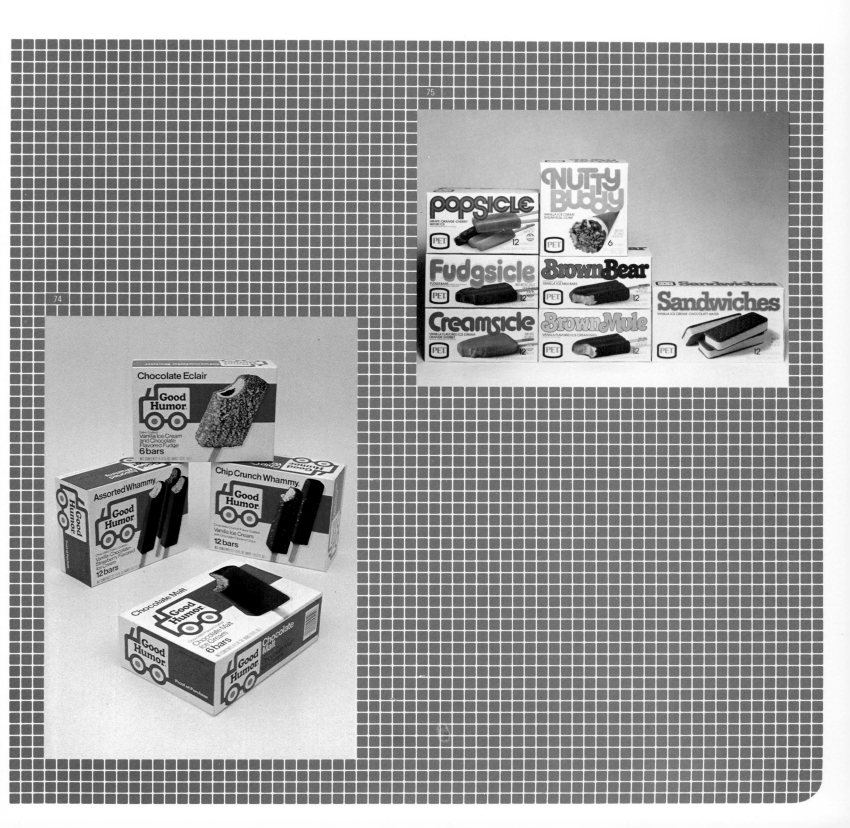

76

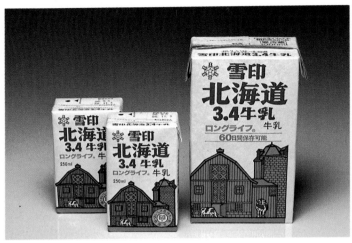

76. **Product:** Long Life Milk
Art Director: Takeo Yao
Design Firm: Yao Design Institute Inc.
Client: Snow Brand

77. **Product:** Eve Yogurt
Art Director: Roger Cayzer
Designer: Roger Cayzer
Design Firm: Young & Rubicam Pty, Adelaide
Client: Dairy Vale

78. **Product:** Neo Soft Margarine
Art Director: Takeo Yao
Design Firm: Yao Design Institute Inc.
Client: Snow Brand

79. **Product:** Friendship Lowfat Cottage Cheese
Design Firm: Martin Jaffe Design Inc.
Client: Friendship Dairies Inc.

80. **Product:** Hokkaido Camembert Cheese
Art Director: Takeo Yao
Design Firm: Yao Design Institute Inc.
Client: Snow Brand

77

78

79

80

81. Product: Hokkaido Butter
    Art Director: Takeo Yao
    Design Firm: Yao Design Institute Inc.
    Client: Snow Brand

82. Product: Pauly Cheese Line
    Art Director: Charles Cybul
    Designer: Brian Rogers
    Design Firm: Source/Inc.
    Client: Swift and Company

83. Product: Danone Yogurt
    Creative Manager: Denis Keller
    Designer: Daniele Matthys
    Design Firm: Cato, Yasumura, Behaeghel
    Client: Gervais-Danone Belgium

84. Product: Tsakudani Jiroubei
    Art Director: Shiro Tazumi
    Designer: Shiro Tazumi
    Client: Jiroubei Inc.

85. Product: Yushintei
    Art Director: Arikuni Tsujimoto
    Designer: Daizen Sugano
    Client: Sanko Inc.

86.	Product:	Hiyamugi Tsuyu
	Art Director:	Masanori Miura
	Designer:	Masanori Miura
	Client:	Ichibiki Inc.

87.	Product:	Shukuhai
	Art Director:	Ritsuji Kametani
	Designer:	Masaaki Hirai
	Client:	Konishi Shuzou Co., Ltd.

88.	Product:	Hipp Baby Food
	Art Director:	John DiGianni
	Creative Director:	John DiGianni
	Design Firm:	Gianninoto Associates, Inc.
	Client:	Hipp Werke (Germany)

89.	Product:	Heinz Baby Cereal
	Design Director:	Ray Perszyk
	Designer:	Bill Johnson
	Design Firm:	Cato Yasumura Behaeghel Inc. (U.S.)
	Client:	Heinz

90.	Product:	Ligne Diet Products
	Creative Manager:	Denis Keller
	Designer:	Francis Colin
	Design Firm:	Cato, Yasumura, Behaeghel
	Client:	Materne Belgium

88

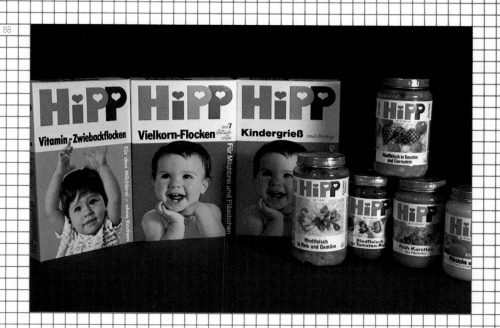

89

90

91. **Product:** Heinz Vinegar
Design Director: Ray Perszyk
Designer: Bill Johnson
Design Firm: Cato, Yasumura, Behaeghel Inc. (U.S.)
Client: Heinz

92. **Product:** Golden Griddle Syrup
Art Director: John Griffin CPC
Creative Director: Irv Koons
Design Firm: Irv Koons Associates, Inc.
Client: C.P.C. International, Inc.

93. **Product:** Heinz Barbecue Sauce
Design Director: Ray Perszyk
Designer: Bob Krusling
Design Firm: Cato Yasumura Behaeghel Inc. (U.S.)
Client: Heinz

94. **Product:** Kraft Buttermilk Dressings
Art Director: Ed Weiss
Designer: Bernard Dolph
Design Firm: Source/Inc.
Client: Kraft, Inc.

95. **Product:** Coleman's Mustard de la Casa
Account Manager: Owen W. Coleman
Designer: Owen W. Coleman
Design Firm: Coleman, LiPuma & Maslow Inc.
Client: Owen & Ellen Coleman

96

96. Product: Wishbone Lite Dressing
    Art Director: John DiGianni
    Creative Director: John DiGianni
    Design Firm: Gianninoto Associates, Inc.
    Client: Thomas J. Lipton, Inc.

97. Product: Bold & Spicy Mustard
    Creative Director: Ronald Peterson
    Designer: Ronald Peterson
    Design Firm: Peterson & Blyth Associates Inc.
    Client: R.T. French

98. Product: Kraft Italian Dressings
    Art Director: Charles Cybul
    Designer: Marybeth Bostram-Cybul
    Design Firm: Source/Inc
    Client: Kraft, Inc.

99. Product: Dundee Marmalades
    Account Manager: Owen W. Coleman
    Designer: Owen W. Coleman/John Rutig
    Design Firm: Coleman, LiPuma & Maslow, Inc.
    Client: The Nestle Company

97

98

99

100. Product: Domino Liquid Brown Sugar
Design Firm: Si Friedman Assoc, Inc.
Client: Amstar

101. Product: Polaner Preserves
Design Director: Joe Selame
Design Firm: Selame Design Staff
Client: M. Polaner & Son

102. Product: Good Seasons Dressing Line
Account Manager: Sal V. LiPuma
Designer: Sal V. LiPuma/Abraham Segal
Design Firm: Coleman, LiPuma & Maslow Inc.
Client: General Foods Corporation

103. Product: Mario Olives
Design Firm: Dixon & Parcels
Client: Shedd's Food div. of Beatrice Foods

104. Product: Dailey's Pickles
Art Director: Mal Feinstein/Paul Gee
Creative Director: Irv Koons
Design Firm: Irv Koons Associates, Inc.
Client: Chas. F. Cates & Sons, Inc.

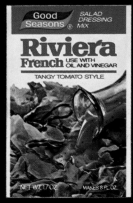

103

104

105

106

107

105. Product: Shiki-no-Sono (Rice Crackers)
     Art Director: Takeo Yao
     Design-Firm: Yao Design Institute Inc.
     Client: Edoichi Sohonpo Inc.

106. Product: Odette/Arcadia Cookies
     Art Director: Nobuyuki Yasuda
     Designer: Nobuyuki Yasuda/Michio Yamato
     Client: Morozoff Co., Ltd.

107. Product: Yo Kam Cookies
     Art Director: Kuniie Misawa
     Designer: Seikin Ishizaki
     Client: Eitaro Sohonpo Inc.

108. Product: Hontagane (Rice Crackers)
     Art Director: Takeo Yao
     Design Firm: Yao Design Institute Inc.
     Client: Edoichi Sohonpo Inc.

108

109. **Product:** Sunshine Graham Crackers
Design Firm: Murtha, DeSola, Finsilver, Fiore, Inc.
Client: Sunshine Biscuits, Inc. (Div. of American Brands)

110. **Product:** Sunshine Cookies
Design Firm: Murtha, DeSola, Finsilver, Fiore, Inc.
Client: Sunshine Biscuits, Inc. (Div. of American Brands)

111. **Product:** Chiparoos Cookies
Design Firm: Murtha, DeSola, Finsilver, Fiore, Inc.
Client: Sunshine Biscuits, Inc. (Div. of American Brands)

112. **Product:** Hand Made Rice Cracker
Art Director: Takeshi Otaka
Designer: Takeshi Otaka
Client: Toyosu Inc.

113. **Product:** Sunshine Cracker/Cookie Line
Design Firm: Murtha, DeSola, Finsilver, Fiore, Inc.
Client: Sunshine Biscuits, Inc. (Div. of American Brands)

<table>
<tr><td>114.</td><td>Product:</td><td>JollyTime Popcorn</td></tr>
<tr><td></td><td>Design Firm:</td><td>Dixon & Parcels Associates, Inc.</td></tr>
<tr><td></td><td>Client:</td><td>American Popcorn Co.</td></tr>
</table>

115. Product: Enduro Snack
Creative Director: Ronald Peterson
Designer: Ronald Peterson
Design Firm: Peterson & Blyth Associates Inc.
Client: American Health

116. Product: Pathmark Cookies
Art Director: Gaylord Adams
Designer: Gaylord Adams
Design Firm: Gaylord Adams & Associates, Inc.
Client: Burry

117. Product: l'ovenbest Donuts
Design Director: Joe Selame
Design Firm: Selame Design Staff
Client: Grand Union

118. Product: Horoku
Art Director: Takeshi Otaka
Designer: Takeshi Otaka
Client: Toyosu Inc.

119

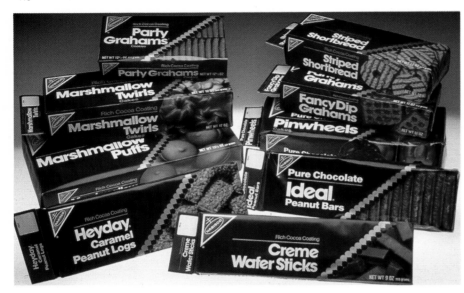

119. Product: Nabisco Chocolate Line of Cookies
     Creative Director: John Lister
     Design Firm: Lister Butler Inc.
     Client: Nabisco Brands

120. Product: Nabisco Gold Line
     Creative Director: John Lister
     Design Firm: Lister Butler Inc.
     Client: Nabisco Brands

121. Product: Nabisco Old Fashioned Crackers
     Creative Director: John Lister
     Design Firm: Lister Butler Inc.
     Client: Nabisco Brands

122. Product: Toastettes
     Art Director: Richard Gerstman
     Designer: Gerstman & Meyers Inc.
     Design Firm: Gerstman & Meyers Inc.
     Client: Nabisco Brands

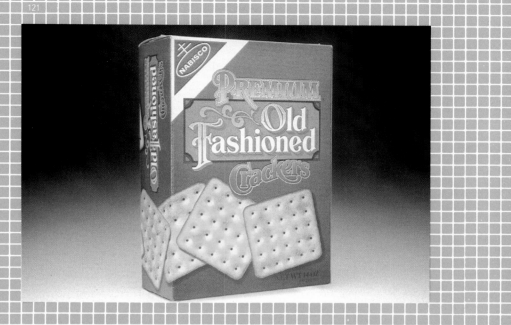

123. Product: Bremner's Jumbo Marshmallow Pies
Art Director: Richard Deardorff
Designer: Laura Hitt
Design Firm: Overlock Howe Consulting Group Inc.
Client: Ralston Purina Co.

124. Product: Bahlsen Pfeffernusser Tin
Design Firm: Babcock & Schmid Associates, Inc.
Client: Bahlsen of America, Inc.

125. Product: Bahlsen Gift Pack
Design Firm: Babcock & Schmid Associates, Inc.
Client: Bahlsen of America, Inc.

126. Product: Shirley Jean Fruit Cake
Graphic Design: Lebanon Packaging Corporation
Client: The Capital Cake Company

127. Product: Burry LU Cookie Line
Account Manager: Owen W. Coleman/Ward M. Hooper
Designer: Ward M. Hooper
Design Firm: Coleman, LiPuma & Maslow Inc.
Client: Burry-LU Inc.

126

127

128

128. Product: Fish Ahoy
     Designer: Susan Healy
     Design Firm: Harte Yamashita & Forest
     Client: Carnation Company

129. Product: Rabbit Chow
     Design Director: Laura Hitt
     Designer: Laura Hitt
     Design Firm: Overlock Howe Consulting Group, Inc.
     Client: Ralston-Purina Co.

130. Product: Farmers Pet Food Range
     Creative Manager: Denis Keller
     Designer: Francis Colin/Erik Vantal
     Design Firm: Cato Yasumura Behaeghel
     Client: Duquesne Purina France

131. Product: Purina Dog Chow
     Creative Director: John Lister
     Designer: Anita Hersh
     Design Firm: Lister Butler Inc.
     Client: Ralston-Purina Co.

132. Product: Puppy Chow
     Design Director: Laura Hitt
     Designer: Laura Hitt
     Design Firm: Overlock Howe Consulting Group, Inc.
     Client: Ralston-Purina Co.

129

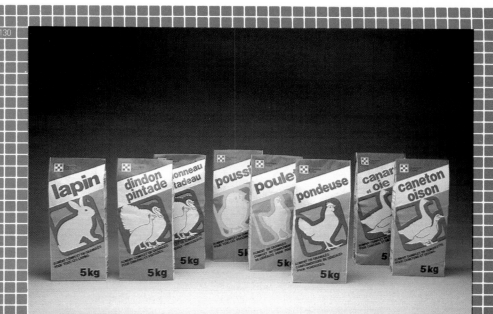

130

131

132

133. **Product:** Australian Dog Chow
     Design Director: Richard Deardorff
     Designer: Laura Hitt
     Design Firm: Overlock Howe Consulting Group, Inc.
     Client: Ralston-Purina International Division

134. **Product:** Alpo/Alamo Dog Food
     Art Director: Donald Flock
     Designer: Donald Flock
     Design Firm: Gaylord Adams & Associates, Inc.
     Client: Allen Products

135. **Product:** Dinner Rounds Dog Food
     Design Director: Clive Chajet
     Design Firm: Chajet Design Group, Inc.
     Client: Campbell's Soup Co.

136. Product: Chuck Wagon Dog Food
     Design Director: Richard Deardorff
     Designer: Laura Hitt
     Design Firm: Overlock Howe Consulting Group, Inc.
     Client: Ralston-Purina Co.

137. Product: Lab Chows
     Design Director: Richard Deardorff
     Designer: Richard Deardorff
     Design Firm: Overlock Howe Consulting Group, Inc.
     Client: Ralston-Purina Co.

138.	Product:	Full Course Dog Food
	Creative Director:	Ronald Peterson
	Designer:	Ronald Peterson
	Design Firm:	Peterson & Blyth Associates, Inc.
	Client:	General Foods Corporation

139.	Product:	Gravy Train (old design)
	Design Firm:	Charles Biondo Design Associates
	Client:	General Foods Corporation

140.	Product:	Gravy Train (redesign)
	Design Firm:	Charles Biondo Design Associates, Inc.
	Client:	General Foods Corporation

141.	Product:	Pet Food
	Designer:	Susan Healy
	Design Firm:	Harte Yamashita & Forest
	Client:	Albers Feeds (div. of Carnation Company)

142.	Product:	Horse Chow/Omolene/Pure Pride Lines
	Design Director:	Richard Deardorff
	Designer:	Richard Deardorff
	Design Firm:	Overlock Howe Consulting Group, Inc.
	Client:	Ralston-Purina Co.

139

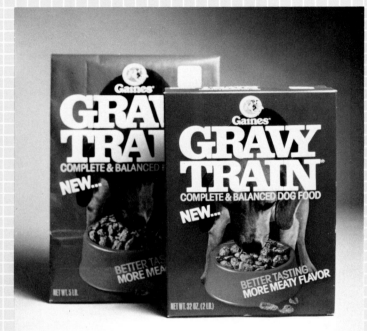

138

140

141

142

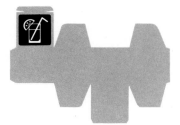

2

Beverage Packages

"Many factors influence people's drinking habits and possibly, the greatest of all these are taste, price and disposable income. When considering affluent, developed, Western countries, disposable incomes are generally sufficiently high to permit taste free rein."

Incpen Discussion Paper No. 7,
"International Drinking Habits"

Learning to like certain beverages is a matter of exposure to them; the customs and social habits of one generation are transmitted to the next. This means that a national tradition or history of a preference for certain drinks is likely to be maintained for as long as national social conditions exist. This is recognized in the folklore and its images of beer for the Germans, wine for the French, soft drinks for the Americans and tea for the English.

Wine

Recessions come and recessions go, but the sales of wine keep on flowing. More than 541 million gallons of imported and domestic vintages were sold in the United States in 1981. These sales figures translate into 2.4 gallons for every American, up from 2.3 gallons a year ago.

Developments abound in U.S. wine packaging—from the rapidly increasing use of the "bag-in-the-box" concept for both bar and home wine packages (introduced in Australia in 1972) to the newly introduced package designs and labels for the "light" and "low-calorie" wines. Also recently available on the U.S. market are individual servings of wine in metal cans. These cans are also available in six-packs for convenience of purchase.

While wine consumption is growing in the United States, France is certainly the

leader with an average consumption of 10.2 gallons per person. For the many Frenchmen who live away from the wine-producing areas, their source for obtaining wine is at the grocer, café and neighborhood wine shop. While most wine sold at liquor retail outlets is in glass bottles, in France, Italy and Spain sizeable quantities are sold in bulk.

Beer

Imported beers are rapidly becoming a factor in the U.S. beer market (4.5 percent in 1981). This has caused many foreign brewers to emulate successfully selling American beer bottle designs. When a Canadian beer was first introduced in the U.S. market a few years ago, the original bottle was short, squat and brown. Shortly thereafter, the beer began appearing instead in slender green bottles, the color and configuration associated by American consumers with premium imported beers.

Other characteristics emerge on a national level. Per capita annual consumption in both the United Kingdom (115 liters) and Germany (147 liters) are high compared to U.S. consumption (80 liters). The preference in Germany is for lager-type beers with high carbonation levels, but in the United Kindom the choice is for less heavily carbonated bitter ales. These differences in national preference have a considerable effect on the overall packaging picture. The bitter flavor of the beer preferred in the United Kingdom is considered enhanced when served on tap, while the German lager is too gaseous to be easily dispensed in this way. The result is that while three-quarters of all United Kingdom beer is served from bulk, over two-thirds of German beer is in consumer packages.

Non-carbonated drinks

The growing popularity of fresh fruit juices has been associated with a move away from glass and cans and toward cartons. Over the past three years consumption of fresh fruit juice and fruit drinks has risen at the rate of 25 percent per year. Most new international brands have been coming in containers of a remarkably uniform appearance ("brik-pak"). Others come in unusually shaped glass bottles.

Coffee

Coffee has long been the leader in novel and attractive packaging. Tinned coffee still accounts for about 80 percent of the total pack in the United States, while in Europe, particularly in Scandinavia, the United Kingdom, Holland, and Germany, the trend is away from cans and in favor of flexible vacuum packages. The annual per capita consumption of coffee in the United States is similar to that of Germany (145 liters) and ahead of France (130 liters) and the United Kingdom (81 liters). Thus, many of the truly important packaging innovations for coffee have been developed in either the U.S. or western Europe.

These developments have included four- and eight-ounce flexible portion-control packages which are gas flushed and then packed in metallized film constructions, as well as one-way valved flexible pouches, composite cans, plastic jars (still in the experimental stage) and flexible one-pound packages.

Carbonated Drinks

Carbonated soft drinks account for about $6 billion annual sales in the United States, where the yearly per capita consumption is 100 liters. Long associated with glass bottles, plastic PET one-liter bottles have now become the accepted mode of packaging. Other innovations include plastic-coated glass bottles, new can shapes, new closures and cans that are less bottom-heavy.

The most obvious technical requirement of a carbonated soft drink package is that the carbonation be retained. In the United Kingdom, where the annual per capita consumption is 32 liters, the carbonation levels are generally high, from two-and-a-half to over four volumes of gas in each volume of drink. In both Germany (51 liters per capita per year) and France (25 liters), carbonation may be below one volume, and PVC bottles can be used.

There are many variables hidden within the apparently simple definition of carbonated soft drinks, and considerable permutations of packaging choice are possible. One example will serve to show that the variations are neither trivial nor independent of national characteristics. The consumption of bitter drinks—tonic water, bitter lemon, ginger ales and soda water—is largely dependent upon the way spirits are drunk. In the United Kingdom, per capita spirit consumption is low, the second lowest among European Economic Community countries, but bitter drinks consumption is high, at 4 liters per head per year. In Germany, spirit consumption is almost twice as much, but annual per capita consumption of bitter drinks amounts to only 0.6 liters. Because of the large soft-drink market, packaging in that product area is more considerably innovative in the United Kingdom than in Germany.

Distilled Spirits

Still the stronghold of the traditional glass bottle, distilled spirits have also been appearing in metal cans, plastic individual-serving bottles and flexible all-plastic pouches. In the United States and Canada, over 1,600 million liters of distilled spirits are consumed annually. White goods (vodka, gin and white rum) have shared in the trend toward lightness and this accounts for the decrease in popularity of brown whiskeys.

144

143

145

143. Product: Elsenham Teas
     Client: Elsenham Quality Foods Ltd.

144. Product: Salada Tea Tin
     Art Director: Ed Morrill
     Designer: Ed Morrill
     Design Firm: Werbin & Morrill, Inc.
     Client: Salada Foods Inc.

145. Product: Royal Tea Bags
     Art Director: John DiGianni
     Creative Director: John DiGianni
     Design Firm: Gianninoto Associates Inc.
     Client: Thomas J. Lipton, Inc.

146. Product: Lipton Herbal Teas
     Art Director: John DiGianni
     Creative Director: John DiGianni
     Design Firm: Gianninoto Associates, Inc.
     Client: Thomas J. Lipton, Inc.

147. Product: Lipton Iced Tea in cans
     Art Director: John DiGianni
     Creative Director: John DiGianni
     Design Firm: Gianninoto Associates, Inc.
     Client: Thomas J. Lipton, Inc.

148. Product: Nestea Iced Tea Line (old design)
     Client: The Nestle Company

149. Product: Nestea Iced Tea Line (redesign)
     Account Manager: Owen W. Coleman
     Designer: Owen W. Coleman/John Rutig
     Design Firm: Coleman, LiPuma & Maslow Inc.
     Client: The Nestle Company

148

149

150. **Product:** Fireside Hot Cocoa Mix
     Designer: Susan Healy
     Design Firm: Harte Yamashita & Forest
     Client: Carnation Company

151. **Product:** Taster's Choice Coffee
     Creative Director: Alvin H. Schechter
     Art Director: Ronald Wong
     Designer: Robert Cruañas
     Design Firm: Schechter Group
     Client: The Nestle Company

152. **Product:** Maxim Freeze Dried Coffee
Art Director: Hajime Tajima
Designer: Riki Watanabe/Teisuke Mura
Client: Ajinomoto General Foods Co., Ltd.

153. **Product:** White Noble Tea
Art Director: Tokihiko Kimata
Designer: Yosei Kawaji/Komei Matsuo
Client: Mitsui Norin Co., Ltd.

154. **Product:** Chat Noir Coffee
Creative Manager: Denis Keller
Designer: Francis Colin
Design Firm: Cato Yasumura Behaeghel
Client: Cafes Chat Noir Belgium

155. **Product:** Chat Noir Coffee
Creative Manager: Denis Keller
Designer: Francis Colin
Design Firm: Cato Yasumura Behaeghel
Client: Cafes Chat Noir Belgium

156. **Product:** New Variety Bleu Rio Coffee
Creative Manager: Denis Keller
Designer: Erik Vantal
Design Firm: Cato Yasumura Behaeghel
Client: Cafes Chat Noir Belgium

152

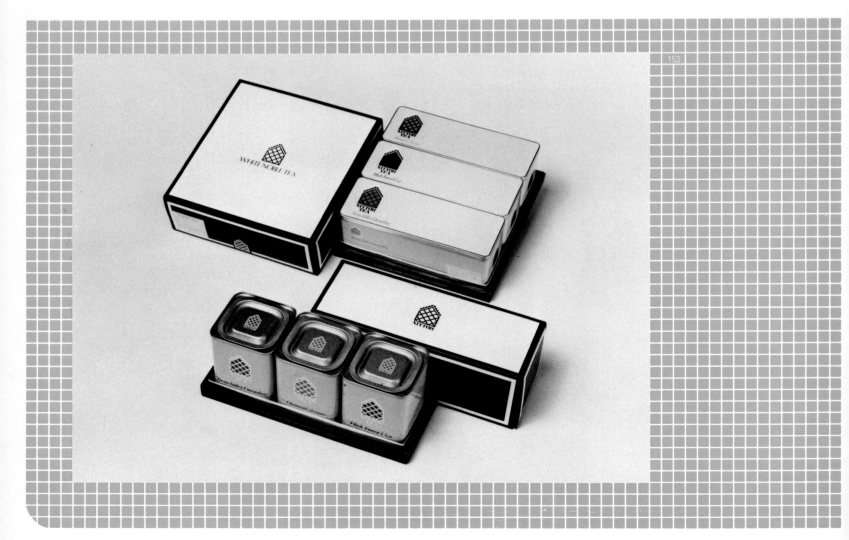

153

154

155

156

157

158

159

157. **Product:** Barrel Head Sugarfree Root Beer
     Art Director: Stuart and Stephen Berni
     Designer: Stuart and Stephen Berni
     Design Firm: Alan Berni Corporation
     Client: Canada Dry Corporation

158. **Product:** Snow Brand Fruit Drink Series
     Art Director: Takeo Yao
     Design Firm: Yao Design Institute Inc.
     Client: Snow Brand

159. **Product:** Barrel Head Sugarfree Root Beer
     Art Director: Stuart and Stephen Berni
     Designer: Stuart and Stephen Berni
     Design Firm: Alan Berni Corporation
     Client: Canada Dry Corporation

160. **Product:** Fruity Fruit Drink Mix
     Art Director: Hajime Tajima
     Designer: Watanabe/Matsuura/Ikeda/Noda
     Client: Ajinomoto General Foods Co., Ltd.

161. **Product:** Ocean Spray Fruit Juices (Aseptic Packages)
     Client: Ocean Spray Cranberries, Inc.

162. Product: Lemon Tree Lemonade
     Art Director: John DiGianni
     Creative Director: John DiGianni
     Design Firm: Gianninoto Associates, Inc.
     Client: Thomas J. Lipton, Inc.

163. Product: Sugar Free Canada Dry Soft Drinks
     Art Director: Stuart and Stephen Berni
     Designer: Stuart and Stephen Berni
     Design Firm: Alan Berni Corporation
     Client: Canada Dry Corporation

164. Product: Canada Dry Soft Drinks
     Art Director: Stuart and Stephen Berni
     Designer: Stuart and Stephen Berni
     Design Firm: Alan Berni Corporation
     Client: Canada Dry Corporation

165. Product: Mello Yello Soft Drink
     Design Director: Clive Chajet
     Design Firm: Chajet Design Group, Inc.
     Client: Coca-Cola Co.

166. Product: Lemon Valley Lemonade
     Design Director: Howard McIlvain
     Designer: Howard McIlvain
     Design Firm: Cato, Yasumura, Behaeghel, Inc. (U.S.)
     Client: Beverage Management

167

168

169

167. Product: Sunkist Orange Soda
     Creative Director: Alvin H. Schechter
     Art Director: Alvin H. Schechter
     Designer: Ronald Wong
     Design Firm: Schechter Group
     Client: Sunkist Soft Drinks, Inc.

168. Product: Royal Crown Decaffeinated Cola
     Art Director: Herbert M. Meyers
     Designer: Lisa Lien
     Design Firm: Gerstman & Meyers Inc.
     Client: Royal Crown Cola Corp.

169. Product: Cerveza Cristal (beer)
     Design Firm: Murtha, DeSola, Finsilver, Fiore Inc.
     Client: Cervecería Nacional S.A.I.C.A.

170. Product: Champale Line
     Art Director: Stuart and Stephen Berni
     Designer: Stuart and Stephen Berni
     Design Firm: Alan Berni Corporation
     Client: Iroquois Brands

170

171

172

173

171. **Product:** Blitz Weinhard Beer Bottle
Art Director: Jerry Andelin
Designer: Primo Angeli
Design Firm: Primo Angeli Graphics
Client: Blitz Weinhard, Pabst Brewing Co.

172. **Product:** Blitz Weinhard Beer Can
Art Director: Jerry Andelin
Designer: Primo Angeli
Design Firm: Primo Angeli Graphics
Client: Blitz Weinhard, Pabst Brewing Co.

173. **Product:** Olympia Brewing Company's Line of Beers
Design Director: Ray Perszyk
Designer: Ray Perszyk/Robert Wile/Jeff Rice
Design Firm: Cato Yasumura Behaeghel Inc. (U.S.)
Client: Olympia Brewing Co.

174. **Product:** Suntory Canned Beer
Art Director: Suntory Design room
Designer: Michio Yamato/Nobuhiro Nakazaki/Akira Kawanishi
Client: Suntory Ltd.

175. **Product:** Schaefer's Cream Ale
Design Director: Clive Chajet
Design Firm: Chajet Design Group Inc.
Client: F & M Schaefer Brewing Co.

176. **Product:** Malta Morena
Creative Director: John Lister
Designer: Anita Hersh
Design Firm: Lister Butler Inc.
Client: Cerveceria Nacional Dominicana C. Por A.
Malta Morena

177. **Product:** Loburg Beer
Creative Manager: Julien Behaeghel
Designer: Denis Keller
Design Firm: Cato Yasumura Behaeghel
Client: Stella Artois Belgium

178. **Product:** Kronenbourg Beer Packs for all European Markets
Creative Manager: Denis Keller
Designer: Christian Callewaert
Design Firm: Cato Yasumura Behaeghel
Client: Kronenbourg Belgium

179. **Product:** Grenzquell Beer
Design Firm: Babcock & Schmid Associates, Inc.
Client: St. Pauli Brewery

180. **Product:** Tribuno Wine Group
Creative Director: James G. Hansen
Art Director: James G. Hansen
Designer: Bernard Dolph
Design Firm: Source/Inc.
Client: The Wine Group

176

177

178

179

180

181

182

181. Product: Vergil Wines
     Art Director: John DiGianni
     Creative Director: John DiGianni
     Design Firm: Gianninoto Associates Inc.

182. Product: Mogen David Wine Group
     Creative Director: James G. Hansen
     Art Director: Bernard Dolph
     Designer: Brian Rogers
     Design Firm: Source/Inc.
     Client: The Wine Group

183. Product: Toko
     Art Director: Shigeichiro Yamamori
     Designer: Kikugoro Mori
     Client: Kojima Sohonten Co., Ltd.

184. Product: Ozeki
     Art Director: Ozeki Shuzo Co. Ltd. Merchandise Planning
                        Committee
     Designer: Iwataro Koike-Kenji Iwasaki
     Client: Ozeki Shuzo Co. Ltd.

185. Product: Reishu Shirayuki
     Art Director: Ritsuji Kametani
     Designer: Tsuyoshi Ikeda
     Client: Konishi Shuzo Co., Ltd.

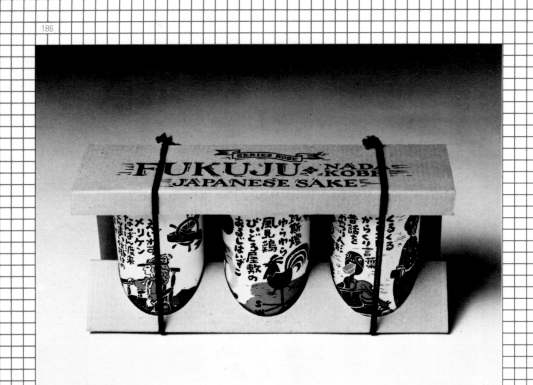

186. **Product:** Fukuju (Woodprint Cup)
     Art Director: Takeshi Ikeda
     Designer: Katsuro Noda
     Client: Fukuju Sake Breweries

187. **Product:** Shogun Wine
     Art Director: Nobuyoshi Ito
     Designer: Nobuyoshi Ito/Takashi Fujita
     Client: Suntory Ltd.

188. **Product:** Jun mai shu
     Art Director: Takeshi Ikeda
     Designer: Katsuro Noda
     Client: Fukujo Sake Breweries Ltd.

189. **Product:** Great Western Wine Line (Champagne)
     Creative Director: Alvin H. Schechter
     Art Director: Ronald Wong
     Designer: Ronald Wong
     Design Firm: Schechter Group
     Client: Wine Spectrum

190. **Product:** California Cellars Wines
     Design Director: Clive Chajet
     Design Firm: Chajet Design Group, Inc.
     Client: The Wine Spectrum div. of the Coca-Cola Co.

189

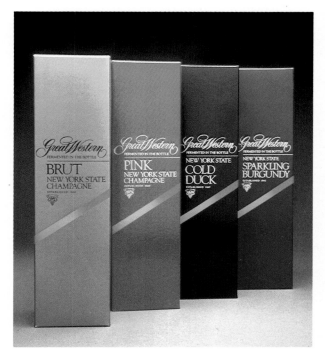

191. **Product:** Botrys Wine Group
Creative Director: James G. Hansen
Art Director: Charles Cybul
Designer: Bernard Dolph
Design Firm: Source/Inc.
Client: The Wine Group

192. **Product:** Mogen David White Cream Concord
Creative Director: James G. Hansen
Art Director: James G. Hansen
Designer: Bernard Dolph
Design Firm: Source/Inc.
Client: The Wine Group

193. **Product:** Great Western Natural Champagne
Creative Director: Alvin H. Schechter
Art Director: Ronald Wong
Designer (carton): Schechter Group Staff
Designer (label): Ronald Wong
Design Firm: Schechter Group
Client: Wine Spectrum

194. **Product:** Dai ho mon
Art Director: Ritsuji Kametani
Designer: Tsuyoshi Ikeda
Client: Konishi Shuzon Co., Ltd.

195. **Product:** Tarukaisen
Art Director: Ritsuji Kametani
Designer: Ritsuji Kametani
Client: Konishi Shuzon Co., Ltd.

193

194

195

196

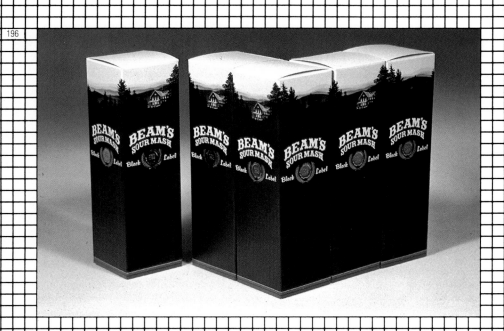

197

198

196. Product: Beam's Sour Mash Box
Creative Director: William J. O'Connor
Art Director: Charles Cybul
Designer: Charles Cybul
Design Firm: Source/Inc.
Client: James B. Beam Distilling Co.

197. Product: Jim Beam's Christmas Box
Creative Director: William J. O'Connor
Art Director: William J. O'Connor
Designer: Bernard Dolph
Design Firm: Source/Inc.
Client: James B. Beam Distilling Co.

198. Product: Suntory Royal
Art Director: Shigetaka Saito
Designer: Shigeru Veki/Hideo Tanigawa/Toshiharu
Amano
Client: Suntory Ltd.

199. Product: Perfect Host Mixes
Design Firm: Charles Biondo Design Associates Inc.
Client: Foremost Foods Co.

200. Product: Jim Beam's American Outpost Box
Creative Director: William J. O'Connor
Art Director: William J. O'Connor
Designer: Charles Cybul
Design Firm: Source/Inc.
Client: James B. Beam Distilling Co.

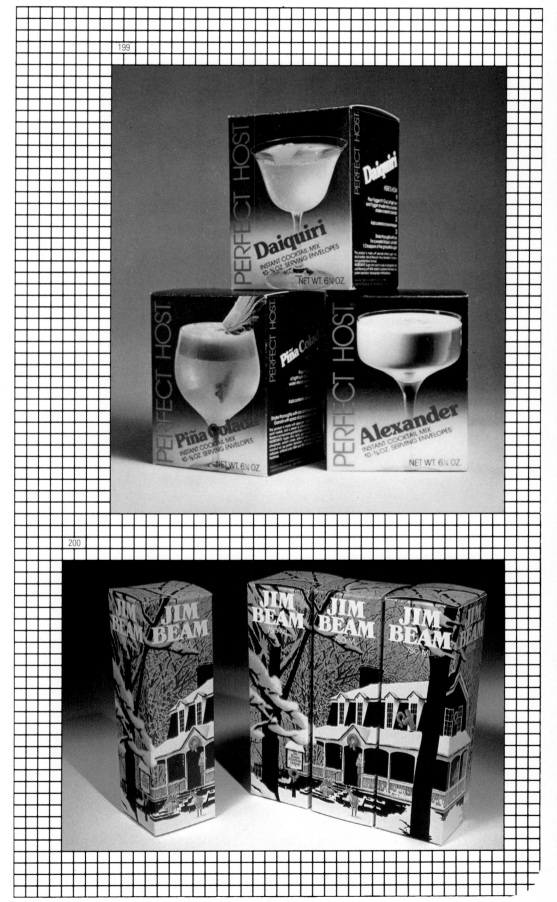

201. Product: Chambord Liqueur Box
     Designer: Steve Hlavna/Norton J. Cooper
     Design Firm: F.N. Burt Company, Inc.
     Client: Charles Jacquin et Cie, Inc.

202. Product: Early Times Whisky
     Creative Director: Ronald Peterson
     Designer: Ronald Peterson
     Design Firm: Peterson & Blyth Associates Inc.
     Client: Brown Forman Distillers

203. Product: Hiram Walker Liqueur Cartons
     Design Firm: Hiram Walker Graphic Design Department
     Client: Hiram Walker & Sons, Inc.

204. Product: Danish Mary Mix
     Designer (graphic): Blum Promotions
     Designer
     (structural):   Mrs. Suvella Richardson, Potlatch
                       Corporation
     Design Firm: Potlatch Corporation
     Client: Somerset Importers, Ltd.

205. Product: DeKuyper Cordial Cartons
     Designer: Kyle Sherwood/Robert Ralston
     Design Firm: Rexham, Warner Packaging Division
     Client: National Distillers

203

204

205

206. **Product:** Coco Lopez
   Art Director: Richard Gerstman
   Designer: Gerstman & Meyers, Inc.
   Design Firm: Gerstman & Meyers, Inc.
   Client: Coco Lopez Imports, Inc.

207. **Product:** Underberg After Drink & Dinner
   Design Director: Robert P. Gersin
   Designer: Georgina Leaf
   Design Firm: Robert P. Gersin Associates Inc.
   Client: Underberg

208. Product: Midori Liqueur
      Art Director: Nobuyoshi Ito
      Designer: Nobuyoshi Ito/Eiko Hirose
      Client: Suntory Ltd.

209. Product: Old Grand Dad Whiskey
      Creative Director: Irving Werbin
      Art Director: Irving Werbin
      Designer: Irving Werbin
      Design Firm: Werbin & Morrill
      Client: National Distillers

210. Product: Parfait Parfait Liqueur Line
     Art Director: NA
     Designer: NA
     Design Firm: NA
     Client: Gilbey Canada, Ltd.

211. Product: Capt. Cook's Cocktail Mixes
     Creative Director: William J. O'Connor
     Art Director: Edward Weiss
     Designer: Bernard Dolph
     Design Firm: Source/Inc.
     Client: James B. Beam Distilling Co.

212. Product: Windsor Canadian Whiskey Box
     Creative Director: Irving Werbin
     Art Director: Irving Werbin
     Designer: Irving Werbin
     Design Firm: Werbin & Morrill Inc.
     Client: National Distillers

213. Product: Suntory Ltd.—The 80th Anniversary Bottle
     Art Director: Nobuyoshi Ito
     Designer: Nobuyoshi Ito
     Client: Suntory Ltd.

214. Product: Juhyo (Suntory Vodka)
     Art Director: Shigeshi Omori
     Designer: Takashi Fujita/Toshiharu Amano
     Client: Suntory Ltd.

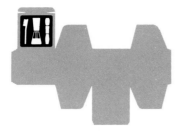

Cosmetics

"Hope is what we sell in cosmetics."

Stephen L. Mayhem
(Voice of the cosmetics
industry for 25 years)

Both demographic and market segmentation studies are critical factors in determining the success of a cosmetic. Different consumers view their cosmetics differently, and what might sell to an affluent consumer is often considered undesirable by a poorer one. This is true for both North American and European consumers.

Similarly, certain products reveal definite national preferences. For example, Germans prefer peppermint toothpaste, while the British show a clear preference for spearmint. Only 38 percent of West German women use lipstick, while 73 percent of Englishwomen do. In England, only 6 percent of the men use any type of toilet water, while in the Netherlands the figure is 12 percent and in France 69 percent.

Mens Toiletries
The 1970s saw the beginning of the trend toward wider use of grooming products by men. Not only are more men using the old standbys like deodorant and aftershave, but they are also beginning to use products that until recently were considered "for women only." Evidence of this growing receptiveness toward new grooming products is shown by the fact that 28 percent of affluent men believe it is appropriate for men to use facial cleansers. Most men are also becoming aware of grooming products through the recommendations of wives or female companions. For men, the scent is the most important factor in selecting a cologne.

Skin Care
Young married women are strongly influenced by the scientific data available on skin care products. This group takes its skin care regimen quite seriously. Hand creams are used by three out of four women. In skin care products, packaging appears to make little difference in the choice of an item, especially among the college educated. Brand loyalty to treatment products runs very high.

Fragrances
Women who buy their own fragrances are far more likely to obtain them in finer department stores than in drugstores. When they purchase a fragrance, it is usually because they have sampled it in the store. When purchasing a fragrance for a woman, more than half the men in the 33–44 age group choose the one suggested by the woman who is to receive the gift.

Personal Care Products
Buying patterns for personal care products vary noticeably with the different educational and income levels of

the consumer. Upscale and downscale consumers have different buying habits. There is also keen interest in having packaging provide more information on personal care products and self-improvement for the consumer.

Affluent Women

The affluent are lured by a quality image, no matter what the price. The association of prestige with a higher priced item is a strong selling point for the upper income consumer. The well-to-do woman is very much interested in acquiring fresh insights on beauty and is willing to try new brands while continuing to support past favorites. Brand name is more important than designer name for this consumer group.

Teen Market

There appears to be a booming teenage market for most cosmetic product categories. Teens are especially brand conscious. The most widely purchased personal care items among teenagers are hair conditioners and nail care products.

Eye, Lip and Nail Care

Young women are strongly influenced by what they see displayed at the makeup counter. Eye shadows are an "impulse" item, and the new shades attract attention. Approximately half the women who use eye shadow wear two shades at once, and will often purchase a complimentary color along with the original choice. The selection of a new set of colors often starts with the lipstick or eye product, rather than nail polish.

Career Women

As women wind their way up the corporate ladder, the use of beauty products helps them feel more confident and more feminine. Most working women buy fragrances and facial products in department stores, hair and nail care products in drugstores. Among career women, cosmetics and fragrances are popular gift items to give and receive, but are rarely exchanged through business.

Color and the Cosmetic Package

Color is the most important element of a package. It is an essential part of the package design and, as the reader can see in the following pages, it enhances even the most pedestrian of cosmetic packages. Our senses are most susceptible to the effects of color, which are capable of evoking a wide range of reactions within us. More than any other factor of package design, color stamps itself on the memory of the consumer to make the package more easily recognizable.

Louis Cheskin* notes many practical examples which emphasize the effect of color on the cosmetic package:

- Sales of one line of shampoos increased when the product was offered in a distinctive brown bottle. In a market where bright colors usually sell, the brown bottle created the line's individual character and made it instantly recognizable by contrast to similar products on the shelves.

- The sales figures for an ivory colored toilet soap which was originally sold in a green wrapper increased when the same soap was put on the market in a blue wrapper.

- An aerosol spray for a cosmetic product was once market tested in a very pale pink package with the brand name in black letters. A series of interviews to test the package's subconscious effect revealed that while the pastel background indicated to the consumer a cosmetic product, the strong contrast of the black lettering evoked the feeling that the product contained in the package was very strong and possibly dangerous.

- A night cream was packaged in a tube as green as an English lawn, while the corresponding day cream was sold in a blue tube. Even after using both creams for quite some time, women constantly confused the two.

- Liquid furniture polish was once sold

in a pink bottle. Market research indicated that many consumers associated the contents of the bottle with some type of beauty product. Others thought of bath salts, baby products and even detergents. Pink, because of its softness and delicacy, belongs to the cosmetics industry. It was used for the furniture polish to suggest a milky, mild product in contrast to the strong products of rival firms.

Psychological Reactions to Color

Red is a powerful color. Physically, it stimulates the digestive system and the circulation of the blood; emotionally, it arouses the forces of self-preservation and signifies strength and virility. In most cases, the use of red must be carefully controlled. Light red is a cheerful color, but dark and bright reds are more likely to induce depression and irritation. Cherry red suggests sensuality.

Orange, more subtle than red, is often used in packages for products associated with physical activity, because it expresses action. It looks clean and appetizing and has the cozy, intimate character of a fire burning in a fireplace.

Yellow, which is cheerful and bright, denotes light, gaiety and warmth. It is especially successful in the Far East. Pale yellow suggests daintiness, golden yellow activity, and a deep strong yellow suggests sensuousness.

Pink suggests femininity and deep affection. While lacking vitality, it gives an impression of gentility and intimacy. A bright magenta pink is associated with frivolity.

Green, quiet and refreshing, is associated with youth, growth and hope. It is undemanding, evoking neither passion nor sadness. (In Arab countries, green is a sacred color, the symbol of the prophets of Islam, and should never be used in package design.) When darkened to olive, it becomes a symbol of decay.

Blue is cool and not subdued. Whereas green suggests the quiet of earth and growing things, blue suggests heavenlike tranquility.

*Louis Cheskin, *How to Predict What People Will Buy* (New York: Liveright Publishing Corp., 1957).

215. **Product:** Carolina Sea Shell Guest Soaps
     Design Firm: Beck Carton Corporation
     Client: Carolina Soap and Candle

216. **Product:** Body Tonics
     Art Director: Seymour M. Kent
     Designer: Neil T. Davis
     Design Firm: Avon Products, Inc.
     Client: Avon Products, Inc.

217. **Product:** Elizabeth Arden Line (Seaqua)
     Creative Director: Ed Morrill
     Art Director: Irving Werbin
     Designer: Ed Morrill
     Design Firm: Werbin & Morrill Inc.
     Client: Elizabeth Arden

218. **Product:** Vitabath
     Creative Director: Ed Morrill
     Art Director: Irving Werbin
     Designer: Irving Werbin
     Design Firm: Werbin & Morrill Inc.
     Client: Beecham, Inc.

217

218

219

220

219. **Product:** Basbon Soap
     Art Director: Yasui Kumai
     Design Firm: Yao Design Institute Inc.
     Client: Shiseido Cosmetics Ltd.

220. **Product:** Christmas Sweater Motif Skin Creme Tubes
     Art Director: Timothy J. Musios
     Designer: Kathryn A. Soghomonian
     Design Firm: Avon Products, Inc.
     Client: Avon Products, Inc.

221. **Product:** Faience
     Designer: Gillian Symonds/Jerry Jankowski
     Design Firm: Southern California Carton Co.
     Client: Lee Pharmaceuticals

222. **Product:** Soft 'n Lovely Bath Beads
     Creative Director: Matin Beck
     Designer: Frank Weitzman
     Design Firm: Gregory Fossella Associates
     Client: Certified Chemical, Inc.

223. **Product:** Spa de Pantene
     Transparency
     Courtesy of: Drug & Cosmetic Magazine

224

225

226

224. **Product:** Rose Milk Lotion
     Design Firm: Si Friedman Associates Inc.
     Client: J.B. Williams Co., Inc.

225. **Product:** Elegance Soap
     Art Director: Hideo Amano
     Designer: Susumu Fukuro
     Client: Kao Soap Co., Ltd.

226. **Product:** Vita Europa Cream Line
     Creative Director: Irving Werbin
     Art Director: Irving Werbin
     Designer: Irving Werbin
     Design Firm: Werbin & Morrill Inc.
     Client: Beecham, Inc.

227. **Product:** Enhance Conditioner
     Creative Director: Dick Young
     Design Firm: Landor Associates, San Francisco
     Client: S.C. Johnson & Son, Inc.

228. **Product:** Whalin' Good Time Shampoo Decanter
     Art Director: Timothy J. Musios
     Designer: Ann M. Beatrice
     Design Firm: Avon Products, Inc.
     Client: Avon Products, Inc.

227

228

229. **Product:** White Rain Hair Spray
Design Director: Ray Perszyk
Designer: Howard McIlvain
Design Firm: Cato Yasumura Behaeghel Inc. (U.S.)
Client: Gillette

230. **Product:** Au Clair Hair Line
Art Director: Hideo Amano
Designer: Noriyuki Inoue/Shigeo Matsuura
Client: Au Clair Cosmetics Co., Ltd.

231. **Product:** Health Works Shampoo
Design Firm: Si Friedman Associates, Inc.
Client: J.B. Williams Co., Inc.

232. **Product:** Blaune Shampoo Line
Art Director: Yukio Kondo
Designer: Kurima Numata/Toshio Taketsuru
Client: Kao Soap Co., Ltd.

233. **Product:** Basbon Shampoo
Art Director: Yasui Kumai
Designer: Takehiko Umekawa/Kazuhiko Adachi/
Minoru Shiokama
Client: Shiseido Co., Ltd.

234

235

236

234. **Product:** Sunnydrop Hair Care Line
Art Director: Tokihiko Kimata
Designer: Yosei Kawaji
Client: Fine Cosmetics Co., Ltd.

235. **Product:** When I Grow Up Shampoo Decanter
Art Director: Timothy J. Musios
Designer: Ann M. Beatrice/Nancy A. Dreisacker
Design Firm: Avon Products, Inc.
Client: Avon Products, Inc.

236. **Product:** Hair Dresser
Art Director: Yukio Kondo
Designer: Kurio Numata/Shigeo Matsuura
Client: Kao Soap Co., Ltd.

237. **Product:** Agree Shampoo
Creative Director: Dick Young
Design Firm: Landor Associates, San Francisco
Client: S.C. Johnson & Son, Inc.

238. **Product:** Enu Shampoo and Rinse
Art Director: Shoichi Hasegawa
Designer: Shoichi Hasegawa
Client: Nakano Seiyaku Co., Ltd.

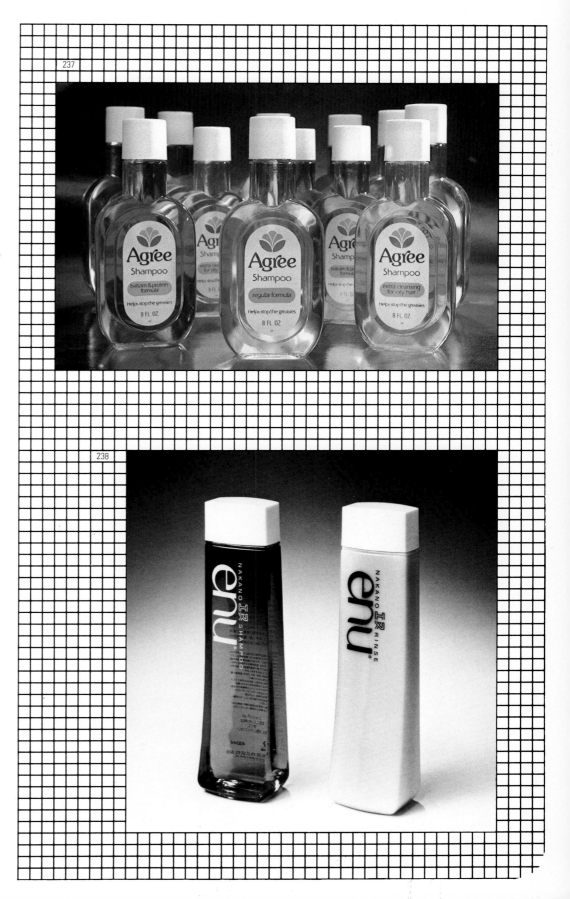

239

239. Product: Moisturizing Cream
Art Director: Seymour M. Kent
Designer: Robert H. Handschuh
Design Firm: Avon Products, Inc.
Client: Avon Products, Inc.

240. Product: Jeunesse Rejuvena
Art Director: Masaru Uda
Designer: Takeshi Ikeda
Client: Momotani Juntenkan Co., Ltd.

241. Product: Mila Cream
Creative Director: Ronald Peterson
Art Director: Ed Kries, Richardson-Vicks Inc.
Designer: Marianne Walther
Design Firm: Peterson & Blyth Associates Inc.
Client: Richardson-Vicks, Inc.

242. Product: Morphee Sundew Series
Art Director: Shoji Tsutsumi
Designer: Akira Takahashi
Client: Kanebo Cosmetics

243. Product: Sahne Cream
Design Firm: Landor Associates, San Francisco
Client: Esai Co., Tokyo

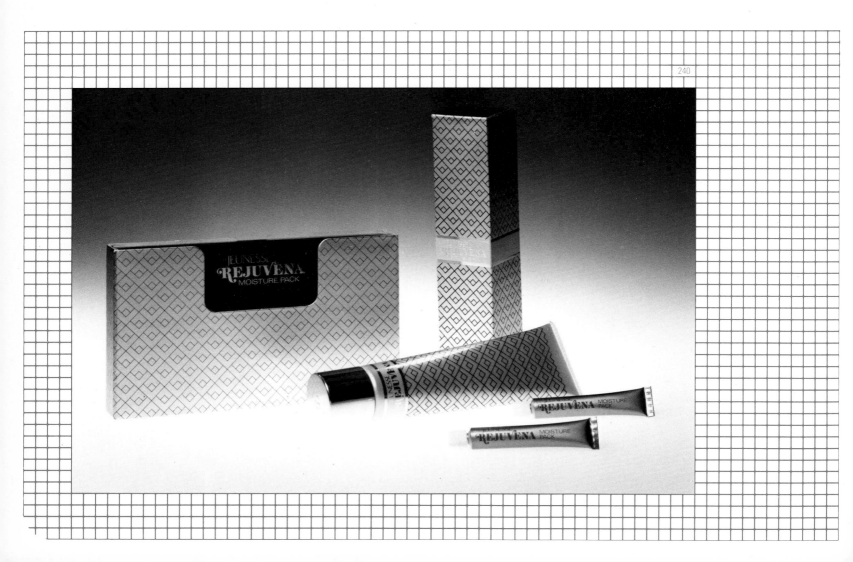

240

241

242

243

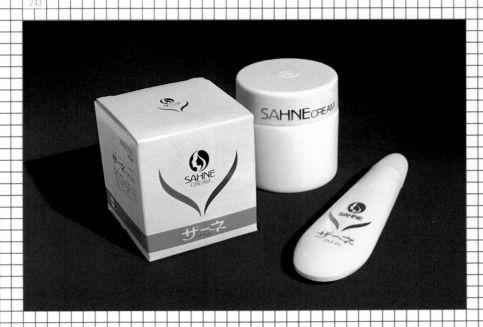

244

245

246

244. Product: Mudd Cleansing Treatment
     Creative Art Director: John S. Blyth
     Designer: John S. Blyth
     Design Firm: Peterson & Blyth Associates, Inc.
     Client: Chattem Drug

245. Product: Rondex Skin Cream for Men
     Art Director: Yuko Hirai/Saburo Nakazaki
     Designer: Yuko Hirai/Ruriko Nomuna/Hiromi Mori
     Client: Olive Manon Cosmetics Inc; Japan Olive Inc.

246. Product: Formula 405 Line
     Art Director: May Bender
     Designer: May Bender
     Design Firm: May Bender Industrial Design
     Client: Doak Pharmacal Company

247. Product: Beauty Facial Cake
     Art Director: Seymour M. Kent
     Designer: Yo Ohama
     Design Firm: Avon Products, Inc.
     Client: Avon Products, Inc.

248. Product: Lindey Skin Line
     Art Director: Yuko Hirai-Japan
     Designer: Yuko Hirai/Ruriko Nomura/Hiromi Mori
     Client: Nippon Relent Cosmetics

247

248

249. **Product:** Lei Shunka, Ekis Skin Line
     Art Director: Yasuhiro Wakutsu
     Designer: Yasuhiro Wakutsu
     Client: Aga Co., Ltd.

250. **Product:** Edge Shave Cream
     Creative Director: Dick Young
     Design Firm: Landor Associates, San Francisco
     Client: S.C. Johnson & Son, Inc.

251. **Product:** Top Condition Line for Men
     Art Director: Seymour M. Kent
     Designer: E. Mark Smith
     Design Firm: Avon Products, Inc.
     Client: Avon Products, Inc.

252. **Product:** Cameron Cologne
     Creative Director: Ben Kotyuk
     Designer: Ben Kotyuk
     Design Firm: Primary Design Group, Inc.
     Client: Kayser Roth

253. **Product:** Matchabelli Cologne and After Shave
     Design Firm: Charles Biondo Design Associates, Inc.
     Client: Matchabelli's Men's Fragrance Line

251

252

253

254. **Product:** Tactics Cologne
  Art Director: Shigeyoshi Aoki
  Designer: Shunsaku Sugiura/Masaki Matsubara
  Client: Shiseido Cosmetics

255. **Product:** Gentlemen's Regiment Line
  Art Director: Timothy J. Musios
  Designer: Richard L. Moyer
  Design Firm: Avon Products, Inc.
  Client: Avon Products, Inc.

256. **Product:** Regatta Cologne
  Design Firm: Si Friedman Associates, Inc.
  Client: J.B. Williams Co., Inc.

257. **Product:** Polo
  Creative Director: Ben Kotyuk
  Designer: Ben Kotyuk
  Design Firm: Primary Design Group, Inc.
  Client: Warner Lauren

256

257

258. **Product:** Pizzazz Makeup Line
Art Director: Seymour M. Kent
Designer: Neil T. Davis
Design Firm: Avon Products, Inc.
Client: Avon Products, Inc.

259. **Product:** Envira Makeup Line
Art Director: Seymour M. Kent
Designer: David P. DiNuccio
Design Firm: Avon Products, Inc.
Client: Avon Products, Inc.

260. **Product:** Oppen Golden Clothes Series
Art Director: Naomoto Takeda
Designer: Kenji Hanamoto
Client: Ryuhodo Seiyaku Co., Ltd.

261. **Product:** Moisture Formula Makeup Line
Creative Director: William J. O'Connor
Art Director: Edward Weiss
Designer: Melissa Woodburn
Design Firm: Source/Inc.
Client: Johnson Products Company, Inc.

262. **Product:** Aha Creation Makeup Line
Art Director: Masahito Matsuda
Designer: Masahito Matsuda/Shonzo Shinoda
Client: Aha Galle Corp.

260

261

262

263. **Product:** Chocolate Cream Lip Gloss
     Creative Director: Ben Kotyuk
     Designer: Ben Kotyuk
     Design Firm: Primary Design Group, Inc.
     Client: Primary Design Group, Inc.

264. **Product:** Moisture Wear Makeup
     Creative Director: Alvin H. Schechter
     Art Director: Ronald Wong
     Designer: Ronald Wong
     Design Firm: Schechter Group
     Client: The Noxell Corporation

265. **Product:** Shiseido Revital Makeup Line
     Art Director: Mr. S. Aoki
     Designer: Mr. S. Sugiura/Mr. P. Togasawa
     Client: Shiseido Cosmetics Co.

266. **Product:** Myrtilla Makeup Line
     Art Director: Tomohiko Nagano
     Designer: Junichi Ishikawa/Masayuki Suzumori
     Client: Pola Cosmetics

267. **Product:** Lady Eighty Makeup Series
     Art Director: Shoji Tsutsumi
     Designer: Akira Takahashi
     Client: Kanebo Cosmetics

268

269

270

271

268. **Product:** Crystalique Line with Cartons
     Art Director: Timothy J. Musios
     Designer: Curtis L. Iverson
     Design Firm: Avon Products, Inc.
     Client: Avon Products, Inc.

269. **Product:** Shiseido Shower Cologne
     Art Director: Mr. S. Aoki
     Designer: Mr. Matsubura/Mr. T. Togasawa
     Client: Shiseido Cosmetics Co.

270. **Product:** Entresse Cologne
     Creative Director: William J. O'Connor
     Art Director: Theodore Hasler
     Designer: Bernard Dolph
     Design Firm: Source/Inc.
     Client: Amway Corporation

271. **Product:** Inoui Perfume
     Design Firm: Murtha, DeSola, Finsilver, Fiore, Inc.
     Client: Shiseido Cosmetics Co.

272. **Product:** His & Hers ASL & EDC Colognes
     Art Director: Seymour M. Kent
     Designer: Robert H. Handschuh
     Design Firm: Avon Products, Inc.
     Client: Avon Products, Inc.

273. **Product:** Albion Perfume Kyoen
     Art Director: Haruyo Tahira
     Designer: Naomi Hosoya
     Client: Albion Cosmetics Inc.

274. **Product:**   Zen/Polo Bottles
Transparency
Courtesy of: Engelhard—Hanouia Liquid Gold
Client: Warner Lauren

275. **Product:**   Eau Givreé Fraiche Cologne
Art Director: Seymour M. Kent
Designer: Robert H. Handschuh
Design Firm: Avon Products, Inc.
Client: Avon Products, Inc.

276. **Product:** Tasha Line
Art Director: Seymour M. Kent
Designer: Arthur R. Torell
Design Firm: Avon Products, Inc.
Client: Avon Products, Inc.

277. **Product:** Toccara Fragrance Line
Art Director: Seymour M. Kent
Designer: Douglas B. Smith/E. Mark Smith
Design Firm: Avon Products Inc.
Client: Avon Products, Inc.

278. **Product:** Epris Perfume Line
Designer (graphic): Bert Pearse
Designer (structural): Steve Scott
Design Firm: Southern California Carton Company
Client: Max Factor

279

280

281

279. Product: Foxfire Line
     Art Director: Seymour M. Kent
     Designer: Douglas B. Smith
     Design Firm: Avon Products, Inc.
     Client: Avon Products, Inc.

280. Product: Auslese Cologne Line
     Art Director: Mr. S. Aoki
     Designer: Mr. S. Sugiura/Mr. Ikeda
     Client: Shiseido Cosmetics Co.

281. Product: Lauren Perfume in Crystal
     Creative Director: Ben Kotyuk
     Designer: Ben Kotyuk
     Design Firm: Primary Design Group, Inc.
     Client: Warner Lauren

282. Product: Tempo Fragrance Line
     Art Director: Seymour M. Kent
     Designer: Douglas B. Smith
     Design Firm: Avon Products, Inc.
     Client: Avon Products, Inc.

283. Product: Senchal Line
     Design Firm: Charles Biondo Design Associates, Inc.
     Client: Charles of the Ritz

282

283

4

Drugs/Health Related Products

"A unified basic line look is a trend that provides good brand identification, ease of manufacture and cost reductions . . . and trends toward simplicity in graphics and appearance and the growing importance of a functional package."

Drug and Cosmetic Industry
report on the Cosmo X Exhibit, June 1982.

Drug packaging differs from other areas of packaging, in that governmental regulations are strictly enforced (especially with pharmaceuticals) and careful attention has to be made as to quantity, quality and safety features of the package.

Package development of pharmaceuticals and over-the-counter (OTC) drugs fall under two categories, package development for new products and modification and execution of an existing package. The second category is an on-going procedure and the needs for change in pharmaceutical packaging can arise from many factors. Some factors (or combinations) are: changing market requirements; change in product formulation; governmental regulations; revision of methods/techniques; technological packaging advances; cost variations; internal recognition of pack shortcomings; market scarcity of packaging components; consumer complaints. This last factor is the greatest generator for change, since the consumer is the ultimate judge in deciding if the package, and the product within the package, will succeed or fail.

The Sexual Revolution
The 1970s has often been characterized

as a "decade of change." Sexual mores were radically altered and many sex-related items were brought out of the closet and onto the shelf.

Sanitary napkins were formerly found in drugstores, where they were carefully wrapped in brown kraft paper to conceal their identity. This product group has recently burst forth in full color and fashion, so that it can actually be placed in most powder rooms. Condom packaging has also been truly revolutionized. From a dark, murky pharmacist's drawer, these products have now moved to self-service racks and supermarket shelves, thereby alleviating the embarassment of having to ask the counterperson for assistance in selection. Featuring soft, sensuous colors, bold graphics, or either photos or silhouettes of both men and women, condom boxes and individual wrappers are designed to appeal to consumers of both sexes.

Over-the-Counter Drugs
Even though the statistics show that as long ago as 1960, ten leading over-the-counter drug items, under watch for some time in the United States, sold more in supermarkets than in their normal retail outlets,[1] more imagination is needed in pharmaceutical design. Since the ethical

[1] James Pilditch, *The Silent Salesman* (Boston: Cahners, 1973).

drugs are selling more in the supermarket aisles, designers must apply "grab me" attention to the package while still keeping a clean, medicinal look. In drugstores, people expect the package to look clean, medicinal and pure. Yet, in the supermarket these products have to be able to stand out and get the consumer's attention. The designer also has to keep in mind that when a person is ill, he is in need of attractive packaging. Why make a sick person even sicker by using an ugly pharmaceutical container?

The 1980s will also see:

- Autoclavable flexible pouches for saline and feeding formulas.
- More unit-dosage packages (solid and liquid).
- Packages capable of holding a one-day's dosage of various medicines in unit-dose form. Already used in Europe, they will soon arrive in the United States.
- Improved visuals on child-resistant plastic closures for bottles. Why are all instructions on an injection molded closure in white, while the arrow or bottle opening feature is also in white?

Prescription Drugs

Prescription pharmaceutical items are generally designed to appeal to the prescribing physician rather than to the general consumer. In this area, line identification is more important than individual product identification—the manufacturer, rather than the individual product, is stressed.

However, because ultimately the prescription drug will be passed on to the patient/consumer, a means should be developed for relabeling the product to make it more easily identifiable. Pharmaceutical packages might be designed with a specific area for the pharmacist's label, with instructions, to be placed. Additionally, there is no reason why the packages cannot be made more esthetically appealing. There is some movement in this area—for example, the attractive compact-like containers for birth control pills and certain types of

powdered potassium supplements that come in brightly colored pouches—but further work in this area is needed.

Repositioning Products

New markets for old products should be reconsidered. It may be possible in certain situations to reach an entirely new market. This has been done successfully in some areas—for example, a product originally marketed as an aid for athlete's foot was repackaged as a relief for so-called "jock itch". Petroleum jelly, used as a general lubricant, could be repositioned as a cosmetic product for nails and rough skin. Talcum powder, long considered a baby or women's product, could be repackaged and more aggressively developed for the men's and athletic markets.

With the increasing popularity of sports and athletics, a number of products could be repositioned. Besides talcum powder, such items as pain liniments, alcohol and witch hazel, skin creams, suntan lotion, elastic bandages, over-the-counter pain killers, eardrops, and tissues might be successfully sold, in redesigned packages with bold, masculine colors and graphics, in this area.

Another trend that is occurring and should be acted upon is the working-mother/stay-at-home father combination. With men becoming more and more involved with childcare responsibilities on a daily basis, fathers are making an increasingly larger proportion of purchases in the areas of disposable diapers, baby creams and lotions, and similar items. New packaging should be designed to reflect this trend; it would not be inappropriate, for example, to use bold, bright colors for baby products instead of the traditional pastels.

The teenage market is another area which certain types of products might be successfully repositioned. Acne preparations, medicated and baby shampoos and soaps, feminine hygiene items and painkillers for menstrual cramps, toothpastes and mouthwashes, sun-protection items and contact lens

accessory products are all candidates for this affluent and free-spending market.

Toothpaste

Toothpaste is certainly one of the most widely used and advertised health-related item, yet there are a number of improvements that can be made in the packaging of this product. One problem with a number of brands is the white cap, which can easily be lost on a white or light-colored sinktop. Another problem is the relative difficulty of getting all the toothpaste out of the tube. Some sort of key device for rolling up the end of the tube would be useful.

One noteworthy change, originating in Germany and Switzerland, is the use of composite flex laminates to replace the flexible metal tubes which are subject to splitting. All-plastic tubes are less successful because of the suck-back effect that occurs in a material that returns immediately to its original shape once the pressure is removed.

In terms of graphics and colors, it is interesting to note that what sells well in Europe is not at all popular in the United States. In Europe, toothpaste designs including pictures or nostalgia-related material are extremely successful. Conversely, American consumers prefer stark, geometric patterns and crisp, primary colors.

Internationalism and Improved Package Design

It is perhaps not surprising to note that Germany and Switzerland, which both have strong pharmaceutical industries, are the leading countries in terms of strong and innovative package design. In general, those countries that want most to develop export markets for their products are those which have superior packaging and package design. The People's Republic of China, which for a very long time had almost no exports on the Western market, is now beginning to show considerable improvement in the packaging of products it exports.

284

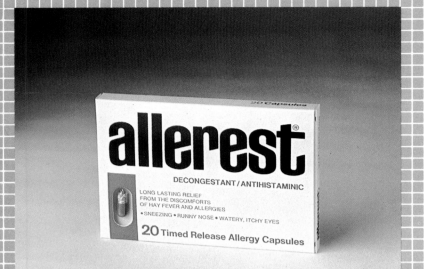

285

286

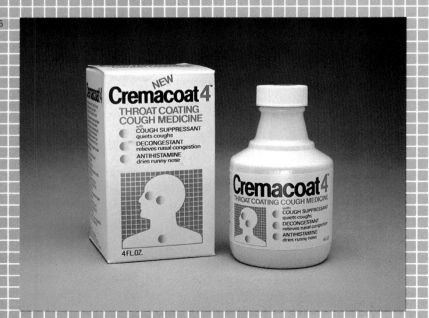

287

284. **Product:** Allerest
     Design Firm: Si Friedman Associates, Inc.
     Client: Pharmacraft

285. **Product:** Vick's Formula 44 & 44D
     Creative Director: Ronald Peterson
     Art Director: Ed Kries, Richardson-Vicks Inc.
     Designer: Lynn Bernick/David Scarlett
     Design Firm: Peterson & Blyth Associates, Inc.
     Client: Richardson-Vicks, Inc.

286. **Product:** Cremacoat
     Creative Director: Ronald Peterson
     Art Director: Ed Kries, Richardson-Vicks, Inc.
     Designer: David Scarlett
     Design Firm: Peterson & Blyth Associates, Inc.
     Client: Richardson-Vicks, Inc.

287. **Product:** Pharma Craft Mist
     Creative Director: John S. Blyth
     Designer: Angele O'Brien
     Design Firm: Peterson & Blyth Associates, Inc.
     Client: Pharmacraft

288. **Product:** Robitussin Night Relief
     Account Director: Owen W. Coleman
     Designer: Owen W. Coleman
     Design Firm: Coleman, LiPuma & Maslow, Inc.
     Client: A.H. Robins Company

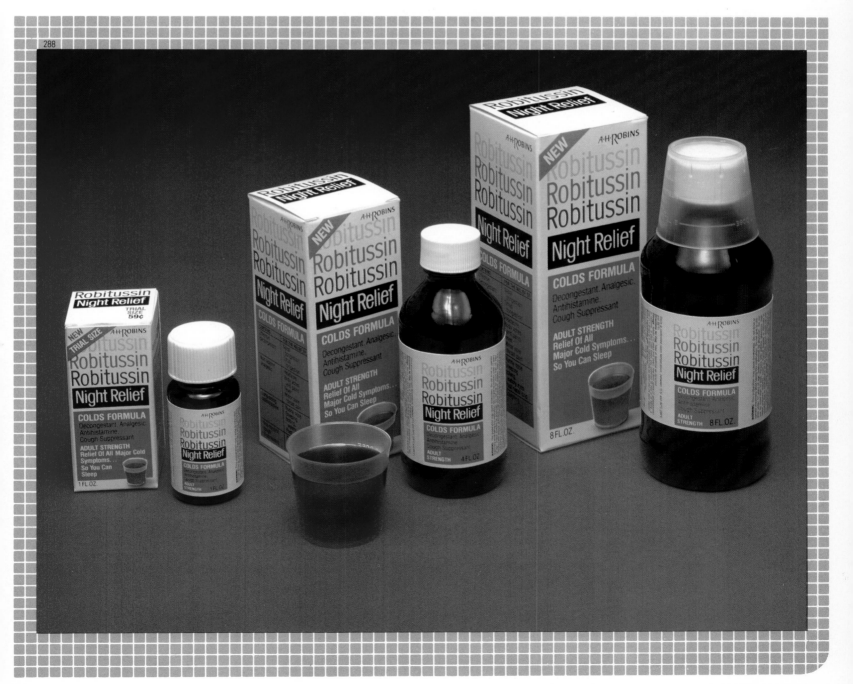

288

289. **Product:** Surround
Creative Director: Ronald Peterson
Art Director: Ed Kries, Richardson-Vicks, Inc.
Designer: Ronald Peterson
Design Firm: Peterson & Blyth Associates, Inc.
Client: Richardson-Vicks, Inc.

290. **Product:** DayCare
Creative Director: Ronald Peterson
Art Director: Ed Kries, Richardson-Vicks, Inc.
Designer: Jim Williams
Design Firm: Peterson & Blyth Associates, Inc.
Client: Richardson-Vicks, Inc.

291. **Product:** Neo Synephrine Nose Drops
Art Director: Soichi Furuta
Designer: Wayne Olsen
Design Firm: Stuart/Gunn & Furuta, Inc.
Client: Winthrop Laboratories

292. **Product:** Neo Synephrine Nasal Spray
Art Director: Soichi Furuta
Designer: Wayne Olsen
Design Firm: Stuart/Gunn & Furuta
Client: Winthrop Laboratories

293. **Product:** New Neo Synephrinal 12 Hour Capsules
Art Director: Soichi Furuta
Designer: Richard Smith
Design Firm: Stuart/Gunn & Furuta, Inc.
Client: Winthrop Laboratories

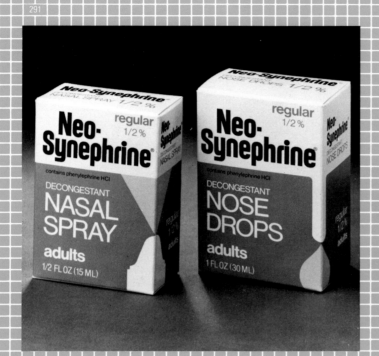

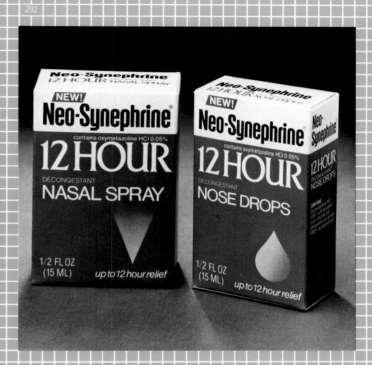

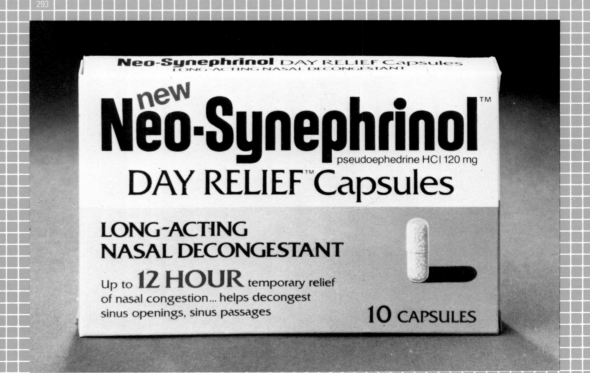

294. **Product:** Remedex
Creative Director: Martin Beck
Designer: Uldis Purins/Michael Catacchio
Design Firm: Gregory Fossella Associates
Client: Sterling Drug

295. **Product:** Caldecort Cream
Design Firm: Si Friedman Associates, Inc.
Client: Pharmacraft

296. **Product:** Biactrin Face Wash
Creative Director: Ronald Peterson
Art Director: Ed Kries, Richardson-Vicks, Inc.
Designer: Lynn Bernick
Design Firm: Peterson & Blyth Associates, Inc.
Client: Richardson-Vicks, Inc.

297. **Product:** ZBT Baby Powder
Creative Director: Irving Werbin
Art Director: Irving Werbin
Designer: Irving Werbin
Design Firm: Werbin & Morrill, Inc.
Client: Glenbrook Labs

298. **Product:** Clearskin 2 Line
Art Director: Seymour M. Kent
Designer: Yo Ohama
Design Firm: Avon Products, Inc.
Client: Avon Products, Inc.

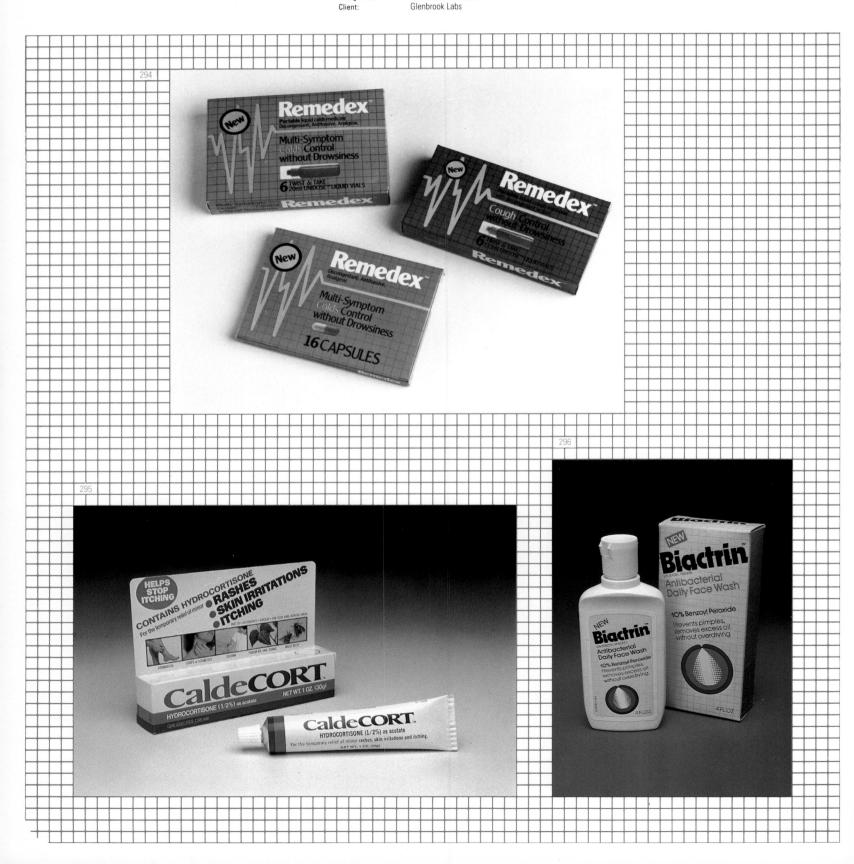

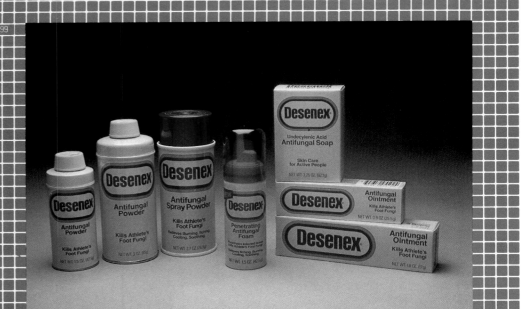

299. **Product:** Desenex Powder
     Creative Director: John S. Blyth
     Designer: Gary Kollberg
     Design Firm: Peterson & Blyth Associates, Inc.
     Client: Pharmacraft

300. **Product:** Aapri Peach Facial Cleanser
     Creative Director: Ben Kotyuk
     Designer: Ben Kotyuk
     Design Firm: Primary Design Group, Inc.
     Client: Gillette

301. **Product:** Dial-a-tan
     Design Firm: Robertz, Webb & Company
     Client: Jovan

302. **Product:** Clearasil Cleanser
     Creative Director: Ronald Peterson
     Art Director: Ed Kries, Richardson-Vicks, Inc.
     Designer: Ronald Peterson
     Design Firm: Peterson & Blyth Associates, Inc.
     Client: Richardson-Vicks, Inc.

303. **Product:** 5 Day Anti-Perspirant
     Design Firm: Si Friedman Associates, Inc.
     Client: J.B. Williams Co., Inc.

304. **Product:** Plenamins Vitamins
Design Director: Richard Howe
Designer: Richard Deardorff
Design Firm: Overlock Howe Consulting Group, Inc.
Client: Rexall Drug Co.

305. **Product:** Health Aids
Design Firm: Yao Design Institute Inc.
Client: Morishita Jintan Co., Ltd.

306. **Product:** Redken Climatress Permanent Wave
Design Firm: Southern California Carton Company
Client: Redken Laboratories, Inc.

307. **Product:** Travel Kit
Designer (graphic): Steve Glick, The Procter & Gamble Co.
Designer (structural): Mrs. Suvella Richardson, Potlatch Corp.
Design Firm: Potlatch Corporation
Client: The Procter & Gamble Co.

308. **Product:** Aloe Vera Gel
Transparency Courtesy of: Drug & Cosmetic Magazine
Client: Fruit of the Earth

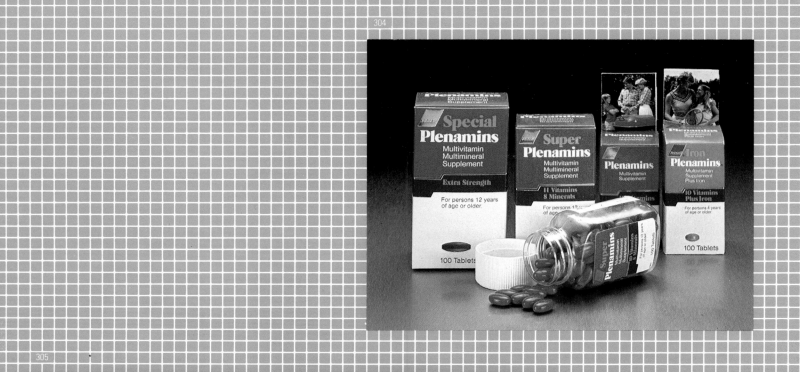

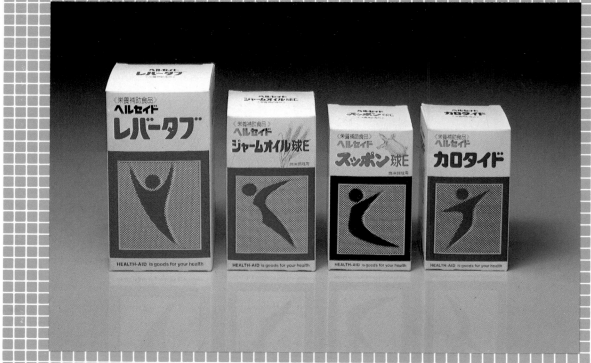

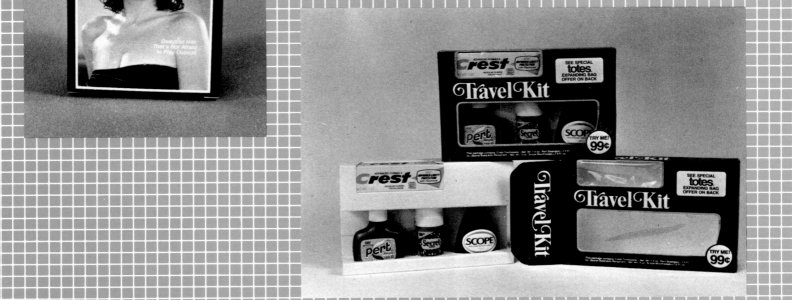

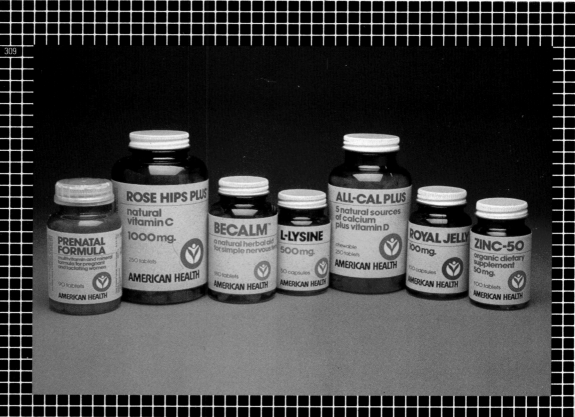

309. **Product:** American Health Vitamin Line
Creative Director: Ronald Peterson
Designer: Ronald Peterson
Design Firm: Peterson & Blyth Associates, Inc.
Client: American Healthaids

310. **Product:** Body Regime Health Aids Line
Design Firm: Babcock & Schmid Associates, Inc.
Client: Scandinavian Health Spa, Inc.

311. **Product:** Vitabank
Design Firm: Si Friedman Associates, Inc.
Client: J.B. Williams Co., Inc.

312. **Product:** American Health Vitamin Line (Aminochel)
Creative Director: Ronald Peterson
Designer: Ronald Peterson
Design Firm: Peterson & Blyth Associates, Inc.
Client: American Healthaids

313. **Product:** Berry C Vitamin Line
Creative Director: Ronald Peterson
Designer: Ronald Peterson
Design Firm: Peterson & Blyth Associates, Inc.
Client: American Healthaids

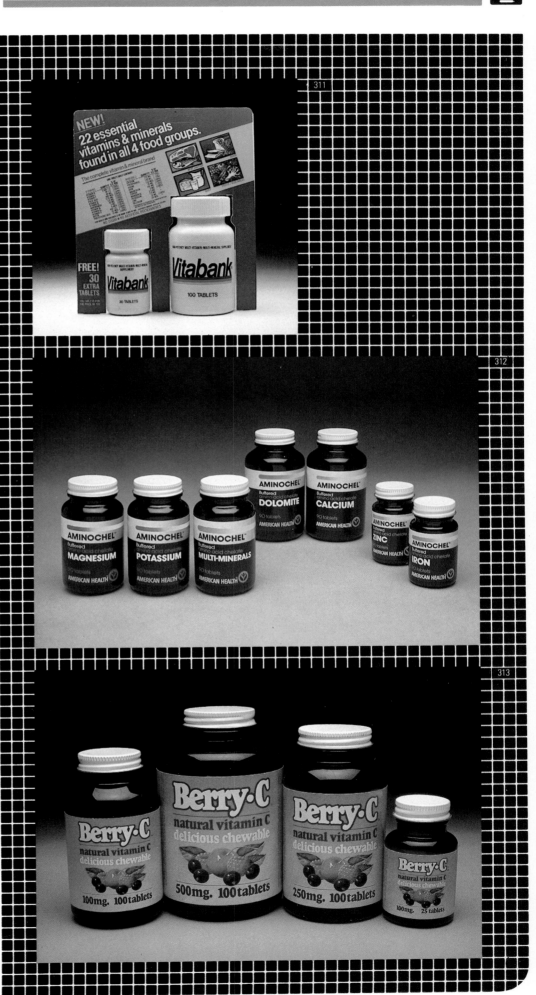

314. **Product:** Tranquility Pads
     Account Director: Owen W. Coleman
     Designer: Owen W. Coleman/John Rutig
     Design Firm: Coleman, LiPuma & Maslow, Inc.
     Client: Principle Business Enterprises

315. **Product:** Cold Wrap
     Art Director: Gaylord Adams/Donald Flock
     Designer: Gaylord Adams
     Design Firm: Gaylord Adams & Associates, Inc.
     Client: Becton Dickinson

316. **Product:** Tempo Antacid
     Designer: Soichi Furuta
     Design Firm: Stuart/Gunn & Furuta, Inc.
     Client: Richardson-Vicks, Inc.

317. **Product:** Phazyme 95
     Art Director: Richard Gerstman
     Designer: Rafael Feliciano
     Design Firm: Gerstman & Meyers, Inc.
     Client: Reed & Carnick Pharmaceuticals

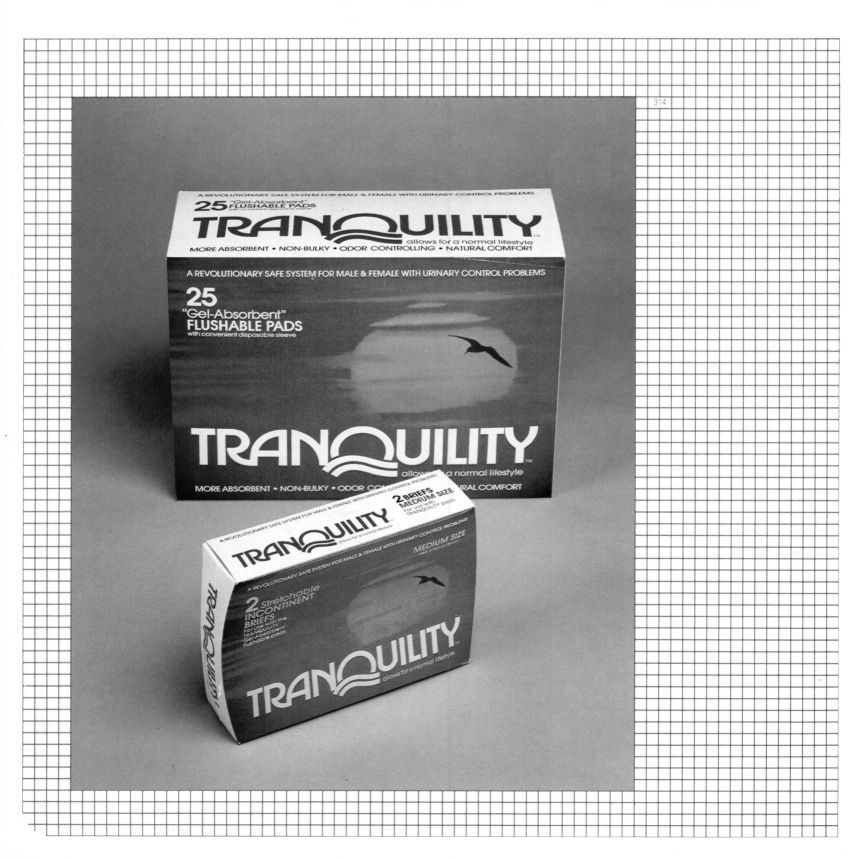

315

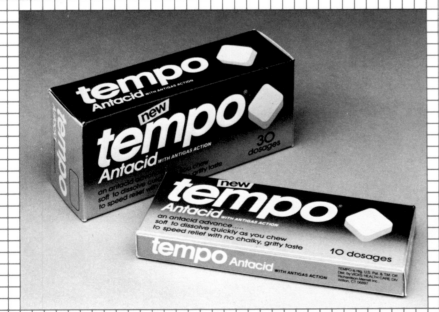
316

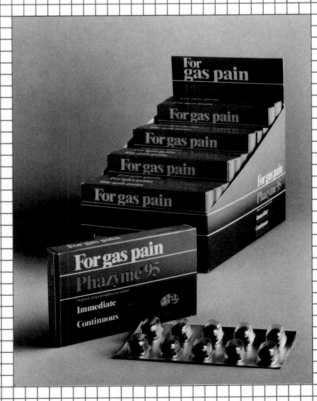
317

318

319

320

318. **Product:** Kleenex 125 Count Tissues
     Creative Director: William J. O'Connor
     Art Director: Edward Weiss
     Designer: Bernard Dolph
     Design Firm: Source/Inc.
     Client: Kimberly-Clark Corporation

319. **Product:** Hoxy Tissues
     Art Director: Katsuhiko Hiramoto
     Designer: Katsuhiko Hiramoto
     Client: Hoxy Co., Ltd.

320. **Product:** Scotties 200 Ply White
     Art Director: Joe Hohmann & Scott Design Staff
     Creative Director: Irv Koons
     Artist: Mal Feinstein/Pat Hallacy/Roger Sainz
     Design Firm: Irv Koons Associates, Inc.
     Client: Scott Paper Co.

321. **Product:** Derma Flex
     Art Director: Norm Weinberger Johnson & Johnson
     Design Firm: Irv Koons Associates, Inc.
     Client: Johnson & Johnson

322. **Product:** Trojans
     Design Firm: Dixon & Parcels Associates Inc.
     Client: Youngs Drug Products

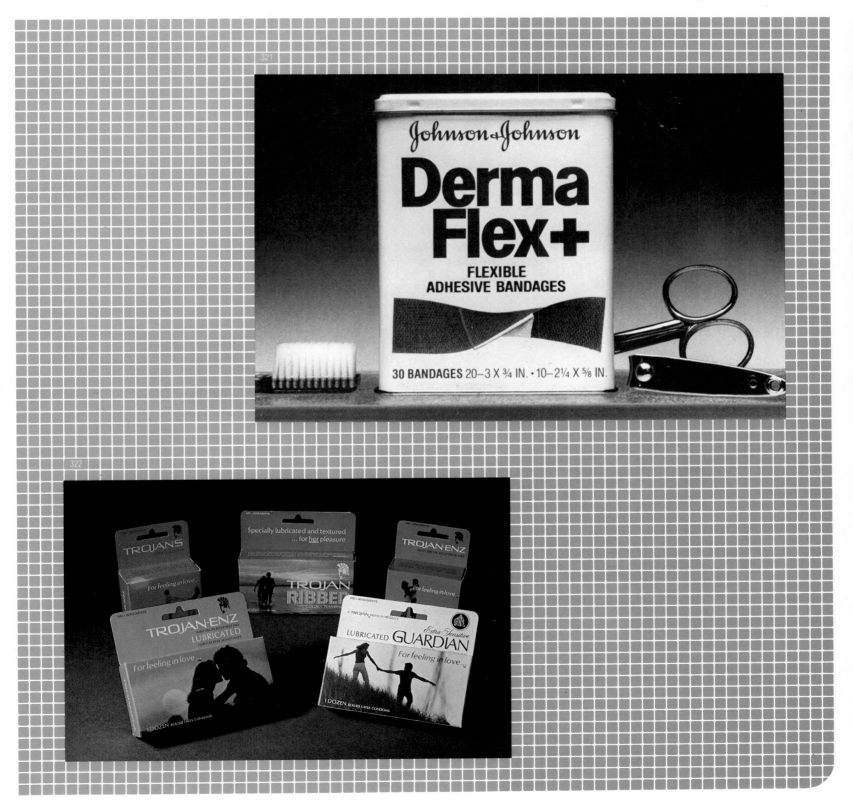

323

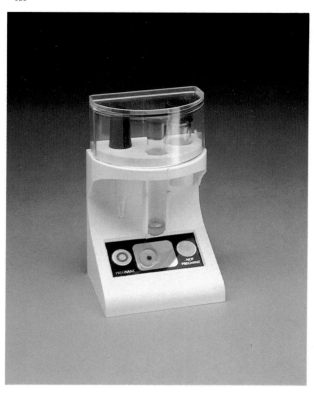

323. **Product:** Pregnancy Test (Acu-Test)
Design Firm: Si Friedman Associates, Inc.
Client: J.B. Williams Co., Inc.

324. **Product:** Acu-Test Pregnancy Test
Design Firm: Si Friedman Associates, Inc.
Client: J.B. Williams Co., Inc.

325. **Product:** Ramses Condoms
Design Firm: Si Friedman Associates, Inc.
Client: Schmidt Laboratories

326. **Product:** Sheik Ribbed Condoms
Design Firm: Si Friedman Associates, Inc.
Client: Schmidt Laboratories

327. **Product:** Zact Lion Dental Cream
Art Director: Masae Fujihashi
Designer: Masae Fujihashi
Client: Lion Co., Ltd.

328

329

330

328. **Product:** Scope Mouthwash
     Designer (graphic): Ray Shaw, The Procter & Gamble Co.
     Designer (structural): Mrs. Suvella Richardson, Potlatch Corp.
     Design Firm: Potlatch Corp.
     Client: The Procter & Gamble Co.

329. **Product:** Mallinckrodt Chemical Line
     Design Director: Richard Deardorff
     Designer: Richard Deardorff
     Design Firm: Overlock Howe Consulting Group, Inc.
     Client: Mallinckrodt, Inc.

330. **Product:** Guard Hello
     Art Director: Tadayoshi Iwatani
     Designer: Kimitoshi Takeda
     Client: Kao Soap Co., Ltd.

331. **Product:** Volumaire Respiratory Exerciser
     Designer (graphic): John Bunn Company
     Designer (structural): Jim Baglio, F.N. Burt Co., Inc.
     Design Firm: F.N. Burt Co., Inc.
     Client: John Bunn Company, Division of Greene & Kellogg, Inc.

332. **Product:** Sominex
     Design Firm: Si Friedman Associates, Inc.
     Client: J.B. Williams Co., Inc.

331

332

Auto/Hardware Packages

"When a Canadian designer created new packaging for EverReady batteries, he made a triangle dominate his design. This, an abstraction of a light beam, was found to have many strong and desirable associations that helped sell the product."

James Pilditch
*The Silent Salesman**

The rapid growth, after many years of neglect, of sophisticated packaging for hardware items is an important factor in the increased sales of these products to a wider consumer market. In *Packaging: The Contemporary Media,* Robert G. Neubauer[†] describes the changes that have been taking place and the improvements these changes have brought about. Formerly, he writes,

> There were a variety of packages for knives and hardware, but display containers, as opposed to individual consumer cartons, prevailed. Hardware was sold from open bins and dark shelves in general stores; the dark surroundings were typical even after the coming of the self-service era. But eventually hardware packaging had to cope with the problems of clerk shortage, pilferage, and impulse buying, and as a result, today's hardware store can compare favorably with brightly lighted supers, department stores, and other sophisticated selling areas. (With the return of the "general store" concept, hardware sections are also to be found in all types of other stores.) Higher profit margins, pegboards, and all the latest packaging techniques characterize hardware merchandizing. The dark, slightly unpleasant aisles of tools and greases are gone for good; they have been replaced by bright packages that prod, promote and sell. Another factor is that we have reverted to doing our own repairs, so that hardware packaging has taken on the total responsibility for instructing us in the work we have to do.

The truth of Neubauer's observations is illustrated with almost casebook clarity in the experience of an old and well established firm of British lock manufacturers. In the past, locks were kept in cabinets behind the counters of most hardware stores. When the customer needed a lock he would have to ask the clerk for assistance in what type of lock to purchase, and then the clerk would have to rummage through various dusty drawers to find the desired lock. Needless to say, this was a highly inefficient use of the clerk's (and the customer's) time. Then, in the late 1960s, the manufacturer developed a self-service merchandising rack that made its locks highly visible and more readily accessible to the consumer.

*James Pilditch, *The Silent Salesman* (Boston, Cahners, 1973).
†Robert G. Neubauer, *Packaging: The Contemporary Media* (New York: Van Nostrand Reinhold, 1973).

In addition to the merchandising rack, the locks were individually blister-packed to keep them new and clean, and were then mounted on cards that showed the various uses and gave general information about each one. The rack and the individual packaging allowed the manufacturer's line of 26 locks to be displayed simultaneously. This highly efficient and up-to-date packaging combination allowed the consumer to see the lock supply available and freed the clerk from having to waste time answering questions and hunting for the desired product.

The increased visibility made a marked increase in sales and this resulted in the manufacturer's ability to reduce the per unit production cost. The reduction, in turn, brought the product to the consumer at a lower price and increased sales for the manufacturer.

It should be readily apparent from the discussion above that packaging for hardware and automotive products—perhaps more than for any other product—does more than merely sell. Most consumers have a pretty good idea, when they go to buy a bar of soap or a frozen pizza, what the product should look like and how it is to be used. This is not necessarily the case with hardware. Hardware, for one, is generally purchased to be used *in conjunction with* another product—a product which may not necessarily be even made by the same manufacturer. A lock, for example, is unlikely to be manufactured by the firm that makes metal cabinets for gymnasium locker rooms. Wrenches are not made by the same firms that manufacture auto parts. Therefore, the package must be not only eye-catching but also highly informative. Size, quantity, compatible equipment, directions for use (often with illustrations) and a number of other pieces of information may be required on the package.

The importance of good display and graphics in impulse buying cannot be stressed too strongly. The psychology of the consumer must be carefully considered. In general, the customer goes to a hardware store seeking a specific item. If the item is sold in a supermarket or super drugstore, the customer has probably not gone there to purchase hardware. In either instance, the impulse purchase will be made only if the product is packaged in such a way as to attract the consumer's attention as an object that may be useful and perhaps even important. This object does not fall into the same category as a new lipstick or another brand of piña colada mix—items that may be wanted but are not necessarily needed. What is being sold is not beauty and romance or a taste of the tropics. The hardware item must be needed, and the package must convey that need in terms of the item's usefulness to the consumer. This may be conveyed through the use of clear and unambiguous images and print information that will prove to the consumer that this is a product that *has to be* purchased.

The use of clear blister packs on hardware and automotive products is probably the most important packaging application in this area. In a do-it-yourself era, the combination of product visibility with well-thought-out instructional drawings or photographs is a boon to both seller and consumer. While not, perhaps, the most glamorous area of package design, hardware packaging does help sell the product and does connote crispness and precision through superior graphics.

Economics of Packaging

Most lightbulbs are sold in carefully (and often ingeniously) designed paperboard sleeves and multi-packs. But within the last few years, a type of generic bulk carton has appeared in various retail outlets, where the bulbs are dump-packed loosely without individual sleeves. The user simply selects the number of bulbs needed, inserts them into a twist-tie polyethylene bag, and has the per-bulb cost totaled at the cash register. The savings to the buyer are considerable (20–30 percent per bulk). Breakage is possible, but with careful handling it is considerably reduced. However, in spite of this successful bulk approach, there still appears to be a need for more inexpensive lightbulb packaging, which currently accounts for about 30 percent of the manufacturers' selling price. Perhaps it is not as exciting as developing a new cigarette or cosmetics package, but it can mean a considerable savings to the manufacturer interested in retaining his premium lightbulb market.

The example involving lightbulbs illustrates one of the most important characteristics of the hardware/auto products market. Even the most progressive and cost-conscious companies have blind spots when it comes to their packaging, often seeing it as a minor detail. Yet the total sum of money spent on packaging materials and labor must equal many of the other important investments made by the company. Time spent studying a packaging operation can be amply repaid.

One preeminent factor is the cost of the pack. The packages illustrated on the following pages are examples of products sold directly to the public, where it has been necessary to create attractive and relatively durable packs. By contrast, in commercial and industrial packaging for a wide range of products, the cost of a pack that will guarantee the undamaged arrival of its contents, irrespective of destination and means of transport, is rarely justified, since most manufacturers include a breakage allowance when shipping.

Industrial packaging is still very mixed in quality. Most firms selling industrial goods, and even those selling consumer durables, have not given much thought to this type of packaging. Even in many companies selling packaged goods, tremendous economies are possible by justification and use of better packaging systems.

333. **Product:** Valvoline Line (old)
Client: Valvoline div. of Ashland Oil Co.

334. **Product:** Valvoline Line (new)
Design Director: Ray Perszyk
Designer: Jim Gobel
Design Firm: Cato, Yasumura, Behaeghel Inc. (U.S.)
Client: Valvoline div. of Ashland Oil Co.

335. Product: Castle Special Motor Oil
     Art Director: Takeo Yao
     Designer: Takeo Yao
     Design Firm: Yao Design Institute Inc.
     Client: Toyota Motor Sales Co., Ltd.

336. Product: Gas Max Line
     Design Director: Richard Howe
     Designer: Richard Deardorff
     Design Firm: Overlock Howe Consulting Group, Inc.
     Client: Gaid Co.

337

338

337. Product: Castle Clean Royal Motor Oil
     Art Director: Takeo Yao
     Designer: Takeo Yao
     Design Firm: Yao Design Institute Inc.
     Client: Toyota Motor Sales Co., Ltd.

338. Product: Casmic Racing Oil
     Art Director: Takeo Yao
     Design Firm: Yao Design Institute, Inc.
     Client: Toyota Motor Sales Co., Ltd.

339. Product: Simoniz Super Poly
     Art Director: Gaylord Adams
     Designer: Gaylord Adams/Donald Flock
     Design Firm: Gaylord Adams & Associates, Inc.
     Client: Union Carbide

340. Product: Castle Long Life Coolant
     Art Director: Takeo Yao
     Design Firm: Yao Design Institute Inc.
     Client: Toyota Motor Sales Co., Ltd.

341. Product: DuPont's Great Reflections Car Wax
     Creative Director: John S. Blyth
     Designer: Penny Johnson
     Design Firm: Peterson & Blyth Associates, Inc.
     Client: Dupont

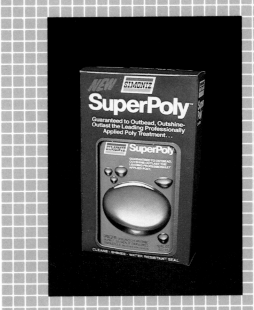

339

340

341

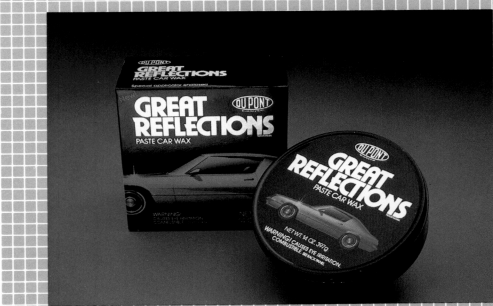

342. **Product:** Casmic Car Line
Art Director: Takeo Yao
Design Firm: Yao Design Institute Inc.
Client: Toyota Motor Sales Co., Ltd.

343. **Product:** Dupont Rain Dance
Art Director: Robert W. Hain
Designer: Jay Robert Wells
Design Firm: Robert Hain Associates, Inc.
Client: E.I. DuPont de Nemours & Co.

344. **Product:** Halogen Headlamps
Designer: General Electric Company
Design Firm: General Electric Company & Robertson Paper Box Co.
Client: General Electric Company

345. **Product:** Pro Fuel Filter Line
Design Firm: Dickens Design Group
Client: C.R. Industries

346. **Product:** Johnson One Touch
Art Director: Misao Hatta
Designer: Hideaki Nishimura/Misao Hatta
Client: Johnson Co., Ltd.

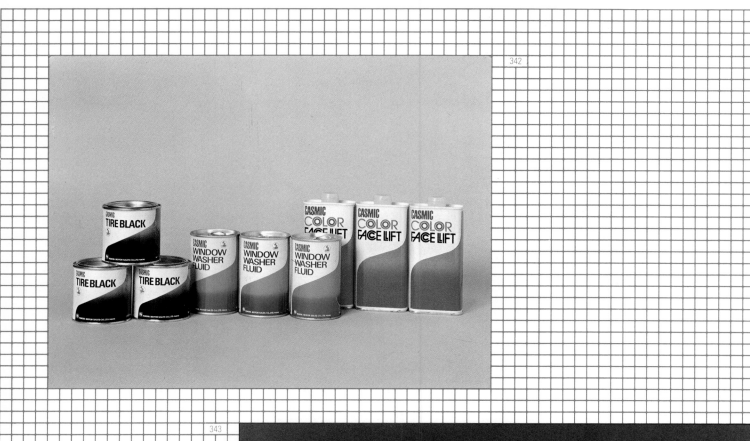

344

345

346

347. **Product:** Sparkleen Pool Chemicals
Designer: Susan Healy
Design Firm: Harte, Yamashita & Forest
Client: GPS Industries

348. **Product:** MAXUM Paint
Creative Director: William J. O'Connor
Art Director: Edward Weiss
Designer: Bernard Dolph
Design Firm: Source/Inc.
Client: Premier Coatings, Inc.

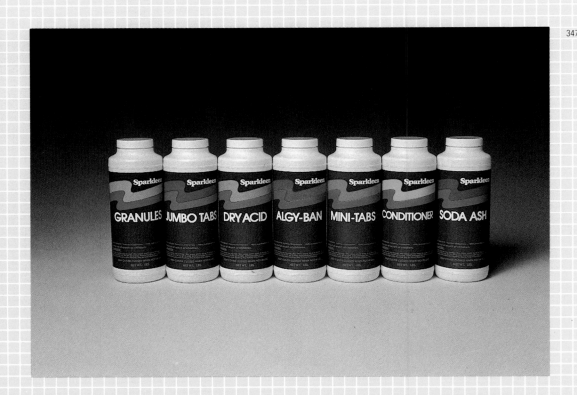

347

348

349. Product: Grease Eater
     Art Director: Gaylord Adams/Donald Flock
     Designer: Gaylord Adams/Donald Flock
     Design Firm: Gaylord Adams & Associates, Inc.
     Client: Union Carbide

350. Product: Garden Hoses
     Art Director: Herbert M. Meyers
     Designer: Rafael Feliciano
     Design Firm: Gerstman & Meyers, Inc.
     Client: Colorite Plastics Company

349

350

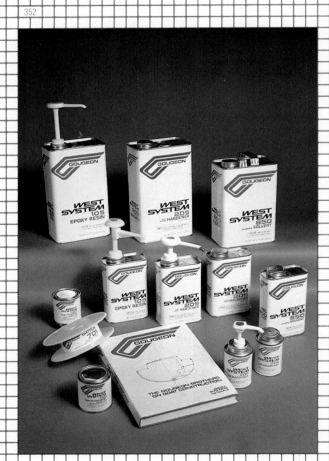

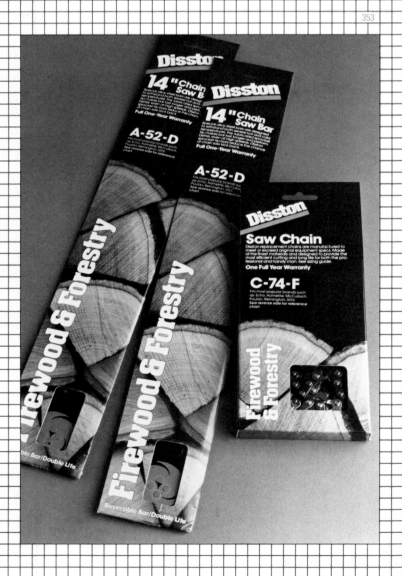

351. **Product:** Pad Painting Kit
     Creative Director: Ed Morrill
     Art Director: Ed Morrill
     Designer: John Kaneleous
     Design Firm: Werbin & Morrill, Inc.
     Client: Paint Master

352. **Product:** Gougeon West System
     Design Firm: Dixon & Parcels Associates, Inc.
     Client: Gougeon Bros.

353. **Product:** Disston Forest Products
     Creative Director: Martin Beck
     Designer: George Boesel
     Design Firm: Gregory Fossella Associates
     Client: Disston

354. **Product:** Sears Craftsmen Tool Set
     Creative Director: Arthur Eilertson
     Designer: George Boesel/Paul Meehan
     Design Firm: Gregory Fossella Associates
     Client: Moore Co.

355. **Product:** Disston Select Handsaw
     Creative Director: Martin Beck
     Designer: John Avery/George Boesel
     Design Firm: Gregory Fossella Associates
     Client: Disston

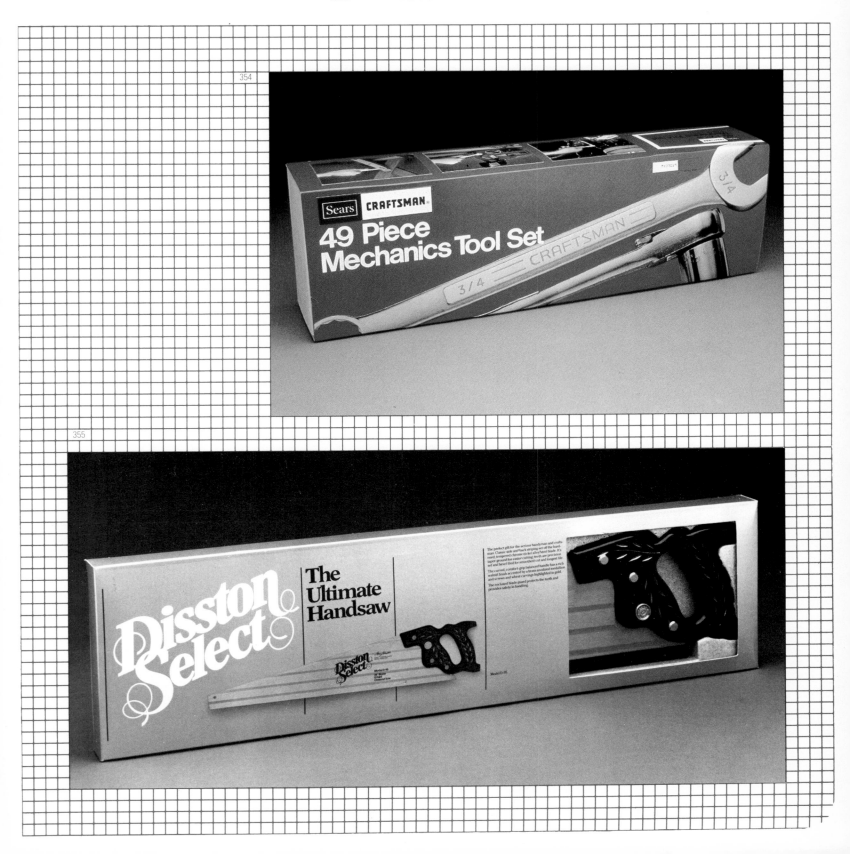

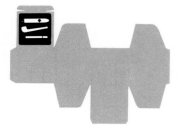

6

Tobacco Products

"In 1958, R.J. Reynolds decided to modernize and clean up the Camel cigarette design. It involved removing two of the three pyramids and doing some things to make it more up-to-date looking. When the first phase, which involved removing one pyramid and changing some lettering was introduced, sales dropped and there was a deluge of customer complaints. The old design was quickly reinstated."

The New York Times
April 14, 1982

Americans spent more than $20 billion in 1980 on tobacco products, with cigars representing a slim $660 million ($40 million of which was spent on smaller or cheaper-priced cigars). An estimated 60 million Americans are considered habitual smokers with each consuming approximately 3,850 cigarettes a year. Health risks appear to have made no impact on the number of smokers or the amount of cigarettes purchased each year. Even though the warning statement is printed on every cigarette pack (sometimes inconspicuously or less dominant in gold ink), cigarette sales are still climbing. In 1964 when the Surgeon General's warning appeared stating that cigarette smoking was definitely hazardous to your health, cigarette sales for that year dropped to 511 billion from 523 billion the previous year. Yet, despite the continuing information and proof that smoking causes lung cancer, in 1975 the figure for cigarette sales jumped to 528 billion.

In tobacco products (as well as in many other product lines) Americans are finally "seeing the light." Although the quest for lightness is not a new concept in the

cigarette world, in 1981–82 some old, well-established cigarette packages were dramatically changed.

When the first light versions of established cigarette brand families came out in the early 1970s, package designers did the obvious—they used white to show lightness (red was equated with full flavor). In 1976, R. J. Reynolds introduced its lowest-tar cigarette, and the color selected by the design firm for this package was silver. Since then, this color has been selected by other low-tar cigarettes, but it now appears that there will be a slow return to more vivid colors in cigarette packaging.

As more products go light, it will become increasingly difficult to differentiate them on a color basis. In fact, with cigarettes it is no longer necessary to emphasize the lightness since every new cigarette product tends to be light. Therefore, the lightness of a cigarette package is not the primary communication concern any longer.

In place of lightness as a primary communication concern, strengthening of brand family identity is taking over in cigarette package design. Many brand families use the same design for all of their cigarette variations, although there may be a different color used for each type. Usually, the entire line of packages will appear together in various forms of advertising to further enhance the feeling of brand family identity.

Cigars

In an effort to boost lagging cigar sales (cigar sales are at their lowest point since 1930, when people first began keeping track) estimated to be down 42 percent from an all time high of 9 billion cigars in 1964, the trend toward "natural" and "light" has finally caught up with this market. In 1980, the consumer saw the introduction by the major cigar manufacturer in the United States of a "deodorized cigar". This cigar looks much like the traditional cigar, but has a filter built into one end, and is specially treated to inhibit the traditional cigar odor. Aimed at luring the young people who cringe at the staid image of the cigar smoker, these new types of cigars are packaged to appeal to this group. The aforementioned deodorized cigar comes in a folding white carton with a natural-colored logo to impart a fresh, clean look to the customer. Whether these new package trends can revive a product that does not appeal to women, is scorned by health groups and is mainly associated with a select group of consumers not in the majority is doubtful. But, the cigar industry does have one bright spot— premium priced cigars (more than 50 cents apiece) are selling strongly, and export sales of elegant hand-rolled cigars rose to 130 million in 1981 from 112 million in 1960.

One major cigar firm has begun testing a new line of cigars on a few selected college campuses on the East Coast. The cigars are slightly crooked cheroots packed in an elliptical tin with an Old West motif stamped on the lid, and the brand name printed on two sides of the can. The main idea behind this different attempt at cigar packaging is to use the current "Western/Cowboy" fad to drum up interest in a lot of men who would not normally smoke a cigar. Another leading tobacco manufacturer is currently marketing a similar product and a new version of the slim "tip" type cigar, those with a built-on holder, in an effort to attract the young smoker.

Pipe Tobacco

The tin and flexible pouch reign supreme in the pipe tobacco field. Many tins are designed to either capture the "imported" look, or to evoke a feeling of nostalgia. The flexible pouches are engineered to give the user a "leather" feel and a "soft" hand. This is usually done by the addition of wax in the structure; this connotes a quality feeling. Many flexible pouches have recloseable features, necessary to preserve the freshness of the product and to keep the tobacco moist until use. And, the flexible pouch is much easier to carry around.

Pipe tobaccos and cigars are relatively staid when compared to cigarettes. Their packaging is designed to give the product either a sophisticated or an all-American look. Even though there has been some attempt to modernize and update cigar and pipe tobacco packaging, improved design is urgently needed to counteract the downward trend in cigar consumption among American smokers.

Chewing Tobacco and Snuff

Of late there has been a small but significant increase in the use of snuff and chewing tobacco among consumer groups that until now were never considered a viable market. The fashion appears to be connected to the rising interest in the American West and Southwest as the inspiration of a number of popular trends.

Packaging for chewing tobacco clearly reflects this regional origin, with Indians and Western themes predominating in design. A certain nostalgia is evident in the graphics, despite the modernization of the resilient containers that exhibit considerably more recent developments in flexible plastics and metals.

The use of snuff, while not increasing as rapidly as chewing tobacco, is nevertheless showing significant gains as flavors such as whiskey or mint are added to the original product. Rigid-package designs reflect the outgoing lifestyle of the consumer markets toward which the product is most successfully aimed.

356. **Product:** Dorados Cigarettes
     Creative Director: John Lister
     Designer: Anita Hersh
     Design Firm: Lister Butler
     Client: Philip Morris International

357. **Product:** Barclay Cigarettes
     Art Director: John DiGianni
     Creative Director: John DiGianni
     Design Firm: Gianninoto Associates Inc.
     Client: Brown & Williamson Tobacco Corporation

358. **Product:** Export A Cigarettes
     Art Director: Gaylord Adams/Donald Flock
     Designer: Gaylord Adams/Donald Flock
     Design Firm: Gaylord Adams & Associates Inc.
     Client: Macdonald Tobacco (Canada)

359. **Product:** Maverick Cigarettes
     Art Director: Alvin Katz
     Designer: Gregory Olanoff
     Design Firm: Luth & Katz, Inc.
     Client: Lorillard

360. **Product:** Sport Suaves
     Art Director: John DiGianni
     Creative Director: John DiGianni
     Design Firm: Gianninoto Associates, Inc.
     Client: Cigarrera la Moderna SA de CV, Mexico

361. **Product:** Kool Ultra
     Art Director: John DiGianni
     Creative Director: John DiGianni
     Design Firm: Gianninoto Associates, Inc.
     Client: Brown & Williamson Tobacco Company

356

357

358

359

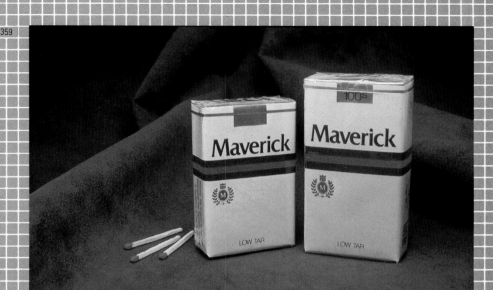

360

361

362

363

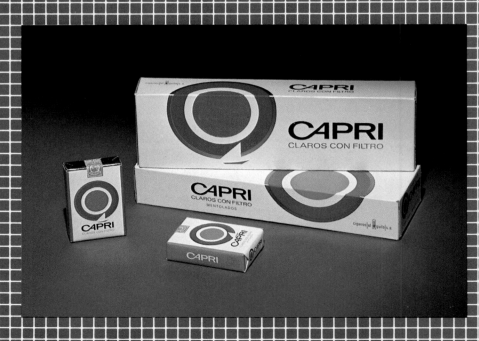

364

362. Product: Tiparillo
Art Director: John DiGianni
Creative Director: John DiGianni
Design Firm: Gianninoto Associates, Inc.
Client: General Cigars

363. Product: Larks
Art Director: John DiGianni
Creative Director: John DiGianni
Design Firm: Gianninoto Associates, Inc.
Client: General Cigars

364. Product: Capri Cigarettes
Art Director: John DiGianni
Creative Director: John DiGianni
Design Firm: Gianninoto Associates, Inc.
Client: Cigarrera la Moderna SA de CV, Mexico

365. Product: Fiesta Cigarettes
Art Director: John DiGianni
Creative Director: John DiGianni
Design Firm: Gianninoto Associates, Inc.
Client: Cigarrera la Moderna SA de CV, Mexico

366. Product: Viceroy Rich Lights
Art Director: John DiGianni
Creative Director: John DiGianni
Design Firm: Gianninoto Associates, Inc.
Client: Brown & Williamson Tobacco Corporation

367. Product: Winston International
Creative Director: Alvin H. Schechter
Art Director: Ronald Wong
Designer: Robert Cruanas
Design Firm: Schechter Group
Client: R.J. Reynolds Tobacco Company

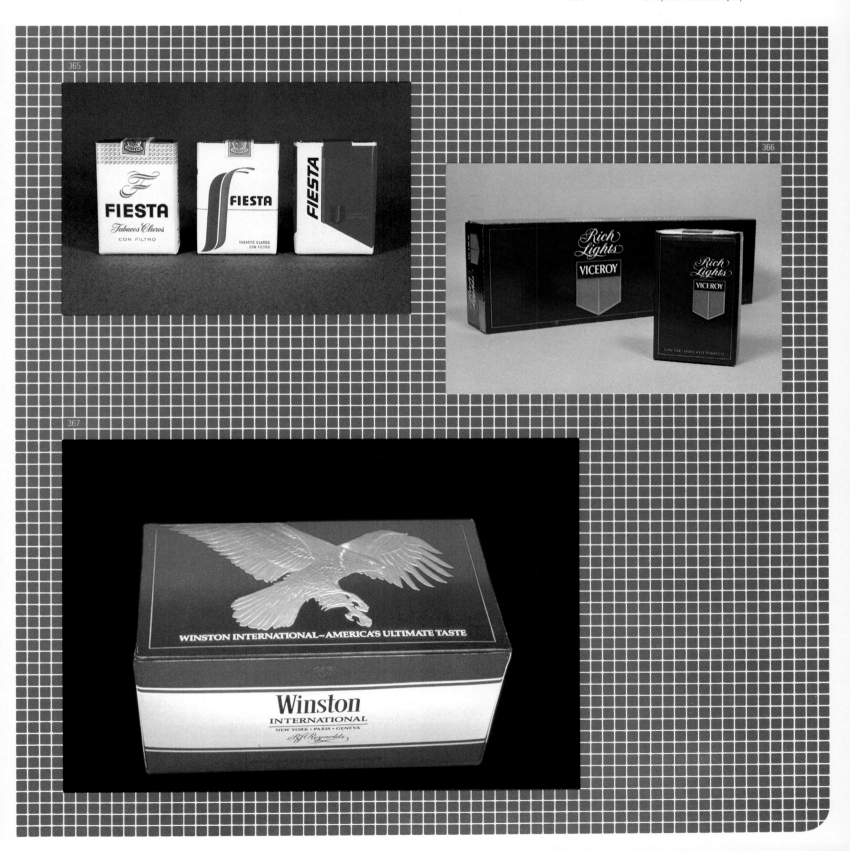

368. Product: Monte Carlo Cigarettes
    Design Firm: Murtha, DeSola, Finsilver, Fiore, Inc.
    Client: Compania Colombiana de Tabaco S.A.

369. Product: Belair
    Art Director: John DiGianni
    Creative Director: John DiGianni
    Design Firm: Gianninoto Associates, Inc.
    Client: Brown & Williamson Tobacco Company

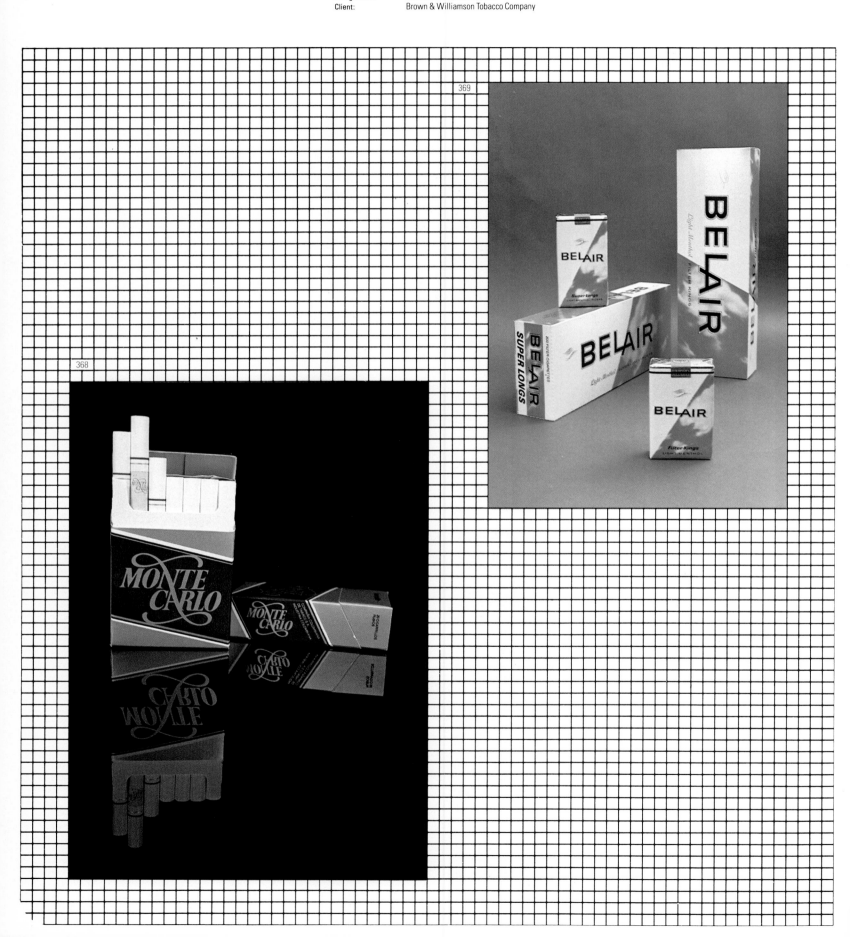

370. **Product:** General Cigar Boxes
Art Director: John DiGianni
Creative Director: John DiGianni
Design Firm: Gianninoto Associates, Inc.
Client: General Cigars

371. **Product:** More Light's 100s
Creative Director: Alvin H. Schechter
Art Director: Ronald Wong
Designer: Ronald Wong
Design Firm: Schechter Group
Client: R.J. Reynolds Tobacco Company

372. **Product:** Export A Lights
Art Director: Gaylord Adams/Donald Flock
Designer: Gaylord Adams/Donald Flock
Design Firm: Gaylord Adams & Associates, Inc.
Client: Macdonald Tobacco (Canada)

373. Product: Marlboro
Art Director: John DiGianni
Creative Director: John DiGianni
Design Firm: Gianninoto Associates, Inc.
Client: Philip Morris

374. Product: Lucky Strike Filters
Art Director: John DiGianni
Creative Director: John DiGianni
Design Firm: Gianninoto Associates, Inc.
Client: Brown & Williamson International Tobacco

375. Product: Bastos Cigarettes
Creative Director: Julien Behaeghel
Designer: Denis Keller
Design Firm: Cato Yasumura Behaeghel
Client: Cinta Belgium

376. Product: Flite Cigars
Creative Director: Irv Koons
Art Director: Irv Koons
Design Firm: Irv Koons Associates, Inc.
Client: Consolidated Cigar Corporation

377. Product: Dudes Cigarettes
Art Director: Gaylord Adams
Designer: Gaylord Adams
Design Firm: Gaylord Adams & Associates, Inc.
Client: Macdonald Tobacco (Canada)

378. Product: Backwoods Cigars
Creative Director: Irv Koons
Art Director: Irv Koons
Design Firm: Irv Koons Associates, Inc.
Client: Consolidated Cigar Corporation

377

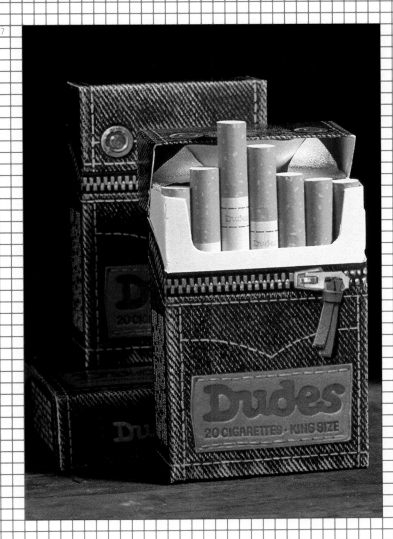

376

378

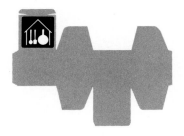

7

Housewares

"The proper role of men and women is not clearly set out by society for all to see. The fact that neither men nor women have a domain clearly and absolutely theirs has created problems."

Anonymous

The Changing Market

For many years, housewares were thought of as strictly a women's market. In the past, this meant appealing to the stereotypical full-time housewife whose main delights in life were centered around the shine on her kitchen floor and the whiteness of her husband's shirts. It was a time when manufacturers and designers thought that merely by appealing to a popular image of housewifely pride, they were giving women what they wanted. It was so easy, then. . . .

The question of what women wanted stumped Sigmund Freud himself, even though he claimed to have spent "thirty years of research into the feminine soul." The full-time, stay-at-home housewife has now become an endangered species. Women now make up 43.5 percent of the American labor force, almost 48 million strong. Nearly half of all the women in the United States are employed, many in demanding careers.

However, the fact that many women work outside of the home does not mitigate the realities of who does the housework—and, perhaps more importantly, who does the household

shopping. Even in most households where the woman holds a full-time job, the majority of housework, including marketing, is done by the woman. Thus, while the stereotypical housewife may be going the way of the ringer-washer, the fact remains that the housewares market is still very much part of the "woman's domain"—changed, perhaps, by differing demands, but nevertheless a part of it.

The Market Breakdown

Professional marketers now subdivide the women's market into four so-called demographic groups: the stay-at-home housewife, the plan-to-work housewife, the "just a job" working woman (at 37 percent, the largest segment), and the career woman. In terms of certain types of housewares purchasing, here is how the market breaks down:

- Women in all four groups buy furniture wax and polish and carpet shampoo, but career women and plan-to-work housewives actually buy more of these products than do stay-at-home housewives.

- The career woman is favored by many product manufacturers because she is likeliest to have the most disposable income. She is also most likely to remain loyal to specific brands—possibly because she has less time to spend investigating products in the supermarket or experimenting with them at home.

- Plan-to-work housewives watch more daytime television than even the stay-at-homes. They have their TV sets on most of the day, and also appear to have their radios on at the same time. Further investigations should reveal what percentage of the advertising messages to which they are subjected is actually absorbed; this may be a factor of whether they are consciously watching or listening to a program or simply utilizing the broadcasts as "background noise."

- In many detergent commercials, the male voice is (somewhat incongruously) used to sell the product. The conventional wisdom is that women respond to the male voice of authority. This may be anachronistic. Needed are more female package designers and other advertising workers who can interpret the female psyche more accurately and develop packages that will appeal to women without insulting their intelligence or common sense.

Newer Markets

Marketing housewares exclusively to the traditional Dick-and-Jane households is an approach that needs serious reevaluation. With changing lifestyles, more than just the image of woman-as-housewife will be affected. With the rising divorce rate and the tendency of both sexes to put off marriage and childbearing to later years, the next decade will see the emergence of a greater market among unmarried adults. Also significant will be the market for people over age 49, who will represent an increasingly greater proportion of the general population.

An extremely new and increasingly substantial market is that composed of so-called "househusbands," that is, stay-at-home men whose wives, either through choice or through necessity, provide the family's financial support. Since this is a clear reversal of the view of housewares as a women's market, it should not be long before significant changes begin to appear not only in product design but in product packaging as well. Darker colors, earth tones and bolder typefaces should become more evident, with the triangular rather than circular design configurations that make the products more appealing to male consumers.

The emergence of all these new markets will surely stimulate the imagination and creativity of those involved in the development and promotion of products aimed toward these individuals. How successfully the industry will react to these new demographic trends remains to be seen.

379

380

381

382

379. **Product:** Jumbo Towel
Art Director: May Bender
Designer: May Bender
Design Firm: May Bender Industrial Design
Client: Hilasal

380. **Product:** Bon-Bory Paper Light
Art Director: Takao Yamada
Designer: Ryo Iwano/Hideo Nakatsuji
Client: Kaizumi Industry Co., Ltd.

381. **Product:** Hefty Bags
Design Firm: Murtha, DeSola, Finsilver, Fiore, Inc.
Client: Mobil Chemical Co.

382. **Product:** Dixie Cups
Art Director: Stuart and Stephen Berni
Designer: Stuart and Stephen Berni
Design Firm: Alan Berni Corporation
Client: American Can Company

383. **Product:** St. Regis Paper Plates
Art Director: Gaylord Adams/Donald Flock
Designer: Gaylord Adams
Design Firm: Gaylord Adams & Associates, Inc.
Client: St. Regis Corporation

383

384. **Product:** Satin Pac Deli Wrap
Designer: Robert Stoming/T. VanderLugt
Design Firm: James River Corporation
Client: James River Corporation

385. **Product:** Kleenex Paper Towels
Creative Director: Irving Werbin
Art Director: Irving Werbin
Designer: JD Grinnell
Design Firm: Werbin & Morrill, Inc.
Client: Kimberly-Clark, Inc.

386. **Product:** Fantastik
Art Director: Alvin Katz
Designer: Erdal Akdag
Design Firm: Luth & Katz, Inc.
Client: Texize

387. **Product:** Rejoice Soap
Designer: American Can Company
Design Firm: American Can Company
Client: Procter & Gamble Co.

388. **Product:** Softsoap
Creative Director: William J. O'Connor
Art Director: William J. O'Connor
Designer: Fred Podjasek
Design Firm: Source/Inc.
Client: Minnetonka, Inc.

384

385

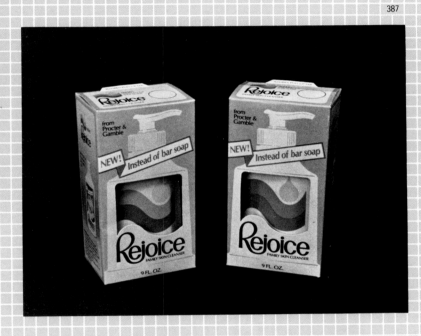

389. **Product:** All
Design Firm: Dixon & Parcels Associates, Inc.
Client: Lever Bros., Co.

390. **Product:** Pinky Detergent
Art Director: Lion Co. Ltd. Package Design Group
Designer: Kiyoshi Takano
Client: Lion Co., Ltd.

391. **Product:** Za Bu Detergent
Art Director: Toshiichi Usui
Designer: Taiichi Fujioka
Client: Kao Soap Co., Ltd.

392. **Product:** Rit Dye Line
Creative Director: Irv Koons
Art Director: John Griffin/CPC
Design Firm: Irv Koons Associates, Inc.
Client: CPC International, Inc.

393. **Product:** Lipasa Sewing Threads
Design Firm: Dil Publicidade Ltda.
Client: Lipasa-Linhas Paulistas

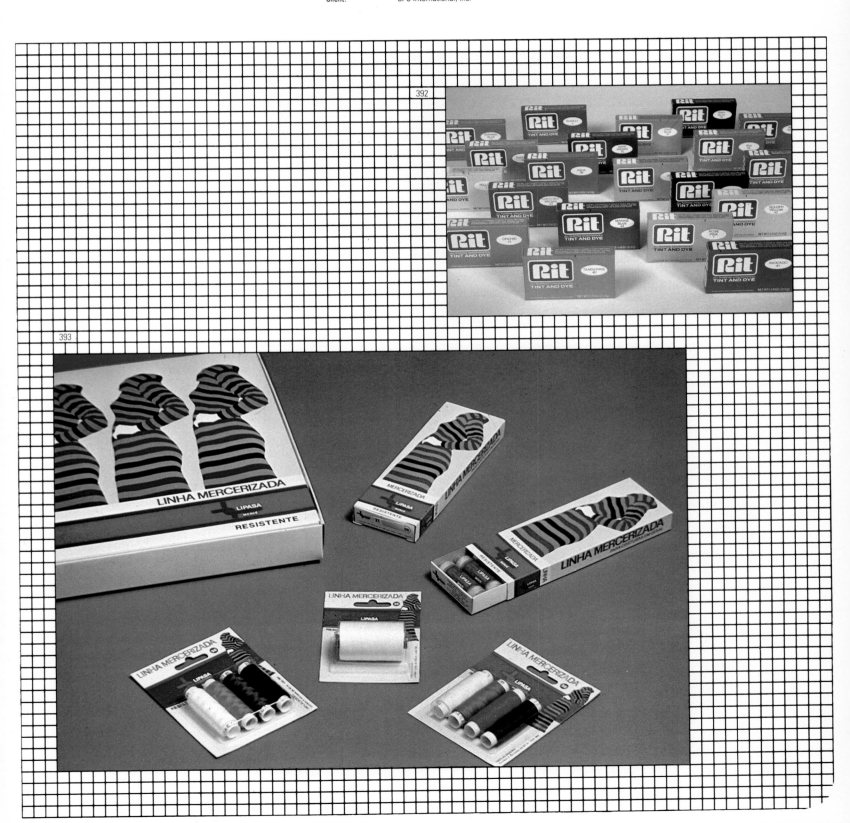

394. **Product:** Rush Floor Cleaner
Creative Director: Denis Keller
Designer: Christian Callewaert
Design Firm: Cato, Yasumura, Behaeghel
Client: Purina Belgium

395. **Product:** Dustbuster
Art Director: Herbert M. Meyers
Designer: Gloria Ruenitz
Design Firm: Gerstman & Meyers, Inc.
Client: Black & Decker Manufacturing Company

396. **Product:** Plant Dolly
Designer (graphic): Don Johnson, Grose, Johnson & Associates
Designer (structural): G. Norman Heaton, Barger Packaging Corp.
Design Firm: Barger Packaging Corporation
Client: Shepherd U.S., Inc.

397. **Product:** Lenor Washing Softener
Creative Director: Tom Salt
Designer: Denis Keller
Design Firm: Cato, Yasumura, Behaeghel
Client: Procter & Gamble Europe

398. **Product:** Drain Power
Design Firm: Charles Biondo Design Associates, Inc.
Client: Glamorene Products

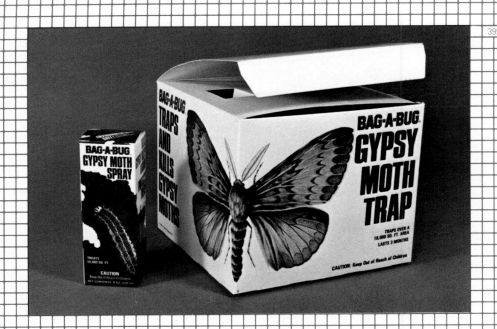

399. Product: Gypsy Moth Trap
Designer (graphic): Stephens Biondi DeCicco
Designer
(structural): Fred Mark, Lebanon Packaging Corp.
Design Firm: Lebanon Packaging Corporation
Client: J.T. Baker Chemical Company

400. Product: 3M Floor Scrubs
Creative Director: Alvin H. Schechter
Art Director: Ronald Wong
Designer: Schechter Group Staff
Design Firm: Schechter Group
Client: 3M Company

401. Product: Door Chime
Designer: Gregory Fossella Associates
Client: General Electric Wiring Division

402. Product: Get 'M Mouse Trap
Design Director: Joe Selame
Design Firm: Selame Design
Client: Wicander Enterprises

403. Product: Scotchgard Carpet Cleaner
Creative Director: Alvin H. Schechter
Art Director: Ronald Wong
Designer: Ronald Wong
Design Firm: Schechter Group
Client: 3M Company

404

404. **Product:** Cowboy Hat Closet Pomander
Art Director: Timothy J. Musios
Designer: Ann M. Beatrice
Design Firm: Avon Products, Inc.
Client: Avon Products, Inc.

405. **Product:** Ecologizer Air Treatment System
Creative Director: Martin Beck
Designer: George Boesel
Design Firm: Gregory Fossella Associates
Client: Rush Hampton Industries

406. **Product:** Pretty as a Picture Air Freshner
Design Firm: Murtha, DeSola, Finsilver, Fiore, Inc.
Client: Lehn & Fink Products Group (Sterling Drug, Inc.)

407. **Product:** Air Salon pas
Art Director: Yasuji Nakagawa
Designer: Yasuji Nakagawa/Katsutoshi Noda
Client: Hisamitsu Seiyaku Co., Ltd.

408. **Product:** Potpourri Garden
Art Director: Leonora Shelsey
Design Firm: Stewart Mosberg Design Associates, Inc.
Client: Cosrich, Inc.

405

409

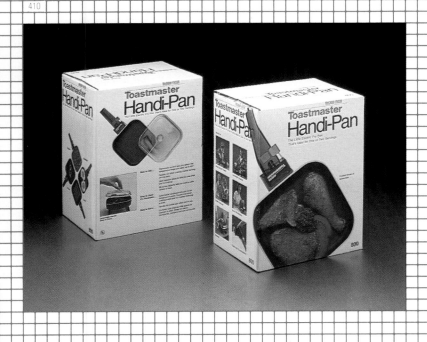

410

411

409. **Product:** Orbit Air Purification System
Creative Director: Martin Beck
Designer: Frank Weitzman
Design Firm: Gregory Fossella Associates
Client: Ion Systems, Inc.

410. **Product:** Handi Pan
Design Director: Richard Howe
Designer: Richard Deardorff
Design Firm: Overlock Howe Consulting Group, Inc.
Client: McGraw Edison

411. **Product:** Show it All
Design Firm: Charles Biondo Design Associates
Client: Corning Glass Works

412. **Product:** National Silverware Containers
Art Director: Stewart Mosberg
Design Firm: Stewart Mosberg Design Associates, Inc.
Client: National Silver Co.

413. **Product:** The Everything Food Processor
Design Firm: Si Friedman Associates, Inc.
Client: Waring

414

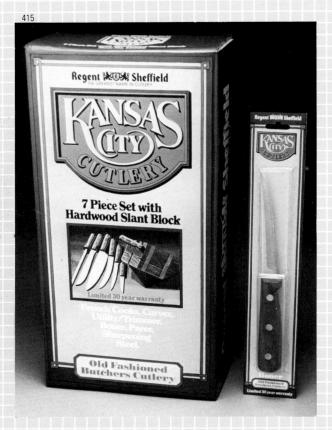

415

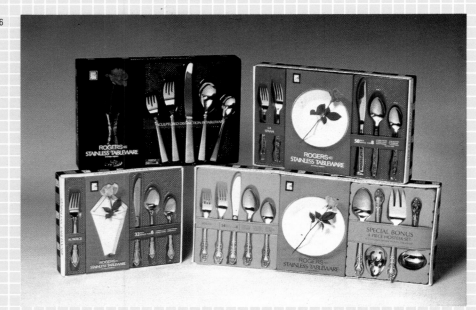

416

414. Product: Rogers Stainless Tableware
    Art Director: Jack Schecterson
    Designer: Jack Schecterson
    Design Firm: Jack Schecterson Associates, Inc.
    Client: Stanley Roberts Inc.

415. Product: Kansas City Cutlery
    Creative Director: Martin Beck
    Designer: Frank Weitzman
    Design Firm: Gregory Fossella Associates
    Client: Regent Sheffield

416. Product: Rogers Stainless Tableware
    Art Director: Jack Schecterson
    Designer: Jack Schecterson
    Design Firm: Jack Schecterson Associates Inc.
    Client: Stanley Roberts Inc.

417. Product: Style Tech Mixing Bowl Set
    Designer: Tets Yamashita
    Design Firm: Harte, Yamashita, & Forest
    Client: Metro Marketing

418. Product: Rogers Excalibur Cutlery Set
    Art Director: Jack Schecterson
    Designer: Jack Schecterson
    Design Firm: Jack Schecterson Associates Inc.
    Client: Stanley Roberts, Inc.

419. Product: Melitta Coffeemakers
    Art Director: Herbert M. Meyers
    Designer: Larry Riddell
    Design Firm: Gerstman & Meyers, Inc.
    Client: Melitta, Inc.

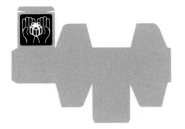

Potpourri

"If an item is given special display, it sells five and a half units for every one sold from its normal shelf position."

Anonymous marketing executive

The following pages vividly illustrate the great variety of packages produced by the package designer. The many different products illustrated are perhaps just as varied as what a package designer is called upon to do on a daily basis. In all these activities, there are certain discrete areas of which the designer must have a clear understanding and in which he or she must be proficient.

The Designer as Graphic Artist

The package designer's proficiency in the graphic arts must extend beyond the area of design itself to all the phases of production—color separation, gravure printing on foil, engraving and printing tolerances and the like. Training in these areas generally comes as a result of graduation from an accredited art or design school combined with extended practical experience.

The Designer as Problem-Solver

The designer must be an efficient planner who can gather all the ingredients and process them into a successful design solution. Since the designer deals with visual communication, he or she must be able to translate thoughts—especially the thoughts of others—into graphic form. He or she must be able to establish realistic goals and objectives, to focus in upon a precise problem within a hazy target and to furnish a sharply defined design solution despite the sometimes nebulous thinking of the client. Unfortunately, the designer cannot afford the luxury of several shotgun solutions; he or she must come up with one rifle-shot design on which the client can rely.

Client Orientation

Any one of several people may be the client—the packer, the package manufacturer, the product manufacturer or the manufacturer's advertising agent. The package designer must be aware of the client's merchandising methods and objectives and must know something about the client's product and brand. Also of vital importance is an understanding of the display characteristics of the intended package.

Consumer Orientation

The ultimate client, of course, is the consumer, for it is he or she who finally pays the bill. Therefore the package designer must be consumer oriented—in tune with the consumer's whims and fancies, reactions and motivations. For it is, finally, at the point of sale that the package design succeeds or fails, as indicated by the almost instantaneous decision of the consumer to buy or not to buy.

Creating Motivation

Today's consumer is primarily female. Women are sometimes said to have at their disposal 85 percent of the nation's wealth (and to influence the disposal of the other 15 percent). Women even buy 62 percent of all hardware items sold at retail, and 80 percent of all the men's shirts sold at Macy's. Thus, catering to the consumer means catering to the woman; it means knowing her reactions and her motivations. When she goes to the supermarket, or to the super-drugstore or to the department store, she is very much aware of brands, products and manufacturers' names. She has gained this awareness from prior experience, from word-of-mouth advertising, from various promotional devices and from radio and TV commercials. But what is really on her mind is her family and her home. When she sees a new food product, she wonders how it will taste to her family, or how she can serve it with other parts of the dinner to make an attractive meal. She is apt to relate a health or beauty product to how it can help her be more attractive or her children better groomed. A new kitchen gadget is related to her kitchen decor or to speeding up the process of food preparation.

The Act of Buying

When the consumer is in the act of buying (the most critical stage for the entire packaging industry), the product, the brand and the manufacturer's image must relate to their uses to the consumer in the consumer's own terms. They must appeal to the buyer in a language that is clearly understood, and in a form that rings the necessary motivational bells to bring the consumer to say, "I'll buy it!"

Integrity

Above all, the package designer must have integrity. Just what this means may best be explained by quoting a brief passage from the American Management Association's book *Packaging is Marketing*, by Dr. Leonard M. Guss: It is a truism that mass production depends upon distribution. Our mass distribution system . . . would be immobilized . . . if the integrity of packaging came so seriously into question that few goods were bought without the consumer's opening the package for inspection. If all consumers insisted on opening and weighing packages . . . the whole system of mass distribution would break down.

Therefore, packaging is an act of faith. It is faith on the part of the seller that he can enclose his product at the factory, . . . transport it, store, it, advertise and sell it, and have it fairly represent him to a buyer who will buy again. It requires an even greater act of faith on the part of the buyer . . . to purchase a product which he cannot see, in a form which must be transmuted to be useful, at a price which he cannot compare in detail, in the expectation that his purchase will satisfy his wants and needs fairly with safety and economy.

This faith is as central to marketing as the faith which underlies credit. When it is betrayed . . . the ability of the economy to act as a mass marketing medium is thereby diminished . . . Every [betrayal] means added cost to the economy in that packaging and other marketing media must work harder to overcome distrust.

420

421

422

420. Product: Parcheesi & Scrabble Games
Art Director: Alvin Katz
Designer: Erdal Akdag
Design Firm: Luth & Katz, Inc.
Client: Selchow & Righter Company

421. Product: Proclaim Box
Design Firm: Boelter Industries, Inc.
Client: Augsburg Publishing House

422. Product: Prang Crayon Case
Art Director: Robert W. Hain
Designer: Jay Robert Wells
Design Firm: Robert Hain Associates, Inc.
Client: American Crayon Co.

423. Product: Jack & Jill and other games
Design Director: Robert P Gersin
Designer: Ron Wong, Kenneth Cook
Design Firm: Robert P Gersin Associates, Inc.
Client: Schaper Manufacturing Co.

424. Product: Cootie Game
Design Director: Robert P Gersin
Designer: Ron Wong/Kenneth Cook
Design Firm: Robert P Gersin Associates, Inc.
Client: Schaper Manufacturing Co.

425. **Product:** Fresh 'n Fancy Cosmetics Kit
Creative Director: Ed Morrill
Art Director: Ed Morrill
Designer: Irving Werbin
Design Firm: Werbin & Morrill, Inc.
Client: Hasbro Industries, Inc.

426. **Product:** Crayola-Rola
Designer (graphic): Janden Hogan, Binney & Smith, Inc.
Designer (structural): Sam Meyers, Central Carton Company
Design Firm: Central Carton Company
Client: Life Lines

427. **Product:** Wiz Wheel
Design Firm: Murtha, DeSola, Finsilver, Fiore, Inc.
Client: Louis Marx Toys

428. **Product:** Capt. Kangaroo Wooden Toys
Creative Director: Ed Morrill
Art Director: Ed Morrill
Designer: Ed Morrill
Design Firm: Werbin & Morrill, Inc.
Client: Hasbro Industries, Inc.

429. **Product:** Stained Glass Kits
Designer: Jack Costa
Design Firm: Taylor Box Company
Client: Aaron Supply Company, Inc.

430. **Product:** Rev 'em Up
Art Director: Jack Schecterson/Barry Herstein
Designer: Jack Schecterson/Barry Herstein
Design Firm: Jack Schecterson Associates, Inc.
Client: Buddy L. Corp.

428

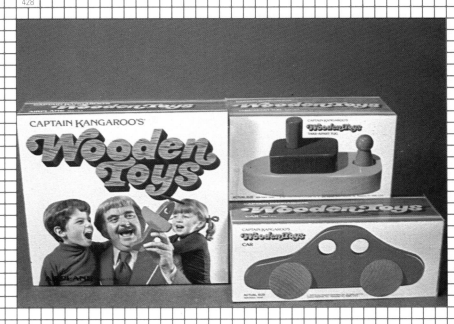

429

430

431

432

433

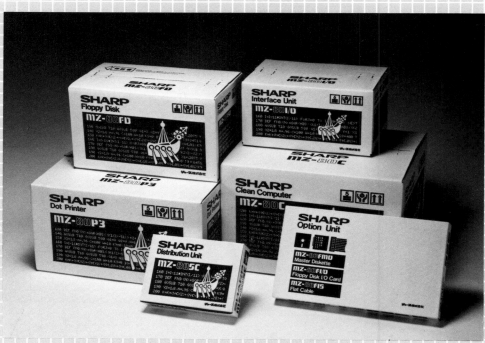

431. **Product:** 3M Scotch Tape
     Creative Director: John S. Blyth
     Designer: Marianne Walther/Judith Miller
     Design Firm: Peterson & Blyth Associates, Inc.
     Client: 3M Company

432. **Product:** Rolodex File
     Designer: Stephen Longo/Philip Lempert
     Design Firm: The Lempert Company
     Client: Rolodex Corporation

433. **Product:** Personal Computer Series
     Art Director: Kiyoshi Sakashita/Koichi Sonoda
     Designer: Koichi Sonoda/Kazuhiko Ashihara
     Client: Sharp Co., Ltd.

434. **Product:** Leroy Controlled Lettering System
     Designer (graphic): Frank Kacmarsky, Joel Lebow Associates
     Designer (structural): Charles Senor, International Folding Paper Box Co., Inc.
     Design Firm: International Folding Paper Box Co., Inc.
     Client: Keuffel & Esser Co.

435. **Product:** Scotch Cassettes
     Creative Director: Julien Behaeghel
     Designer: Christian Callewaert
     Design Firm: Cato, Yasumura, Behaeghel
     Client: 3M Europe

434

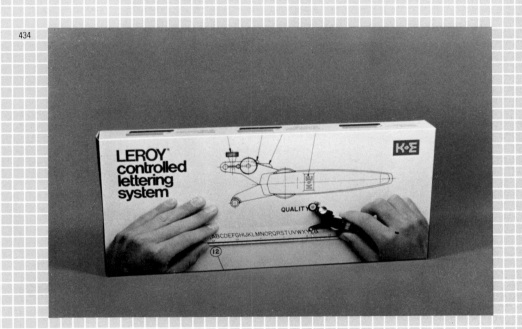

435

436. Product: Vivitar 742 XL Camera
     Designer: Tets Yamashita
     Design Firm: Harte, Yamashita, & Forest
     Client: Vivitar

437. Product: Lenses
     Designer: Tets Yamashita/John Baker
     Design Firm: Harte, Yamshita, & Forest
     Client: Kiron

438. Product: Duotang Report Covers
     Creative Director: Arthur Eilertson
     Designer: Frank Weitzman
     Design Firm: Gregory Fossella Associates
     Client: Schaeffer Eaton div. of Textron

439. Product: Koh-I-Noor Rapidograph Pens
     Art Director: Robert W. Hain/Frederick B. Hadtke
     Designer: Jay Robert Wells
     Design Firm: Robert Hain Associates, Inc.
     Client: Koh-I-Noor Rapidograph

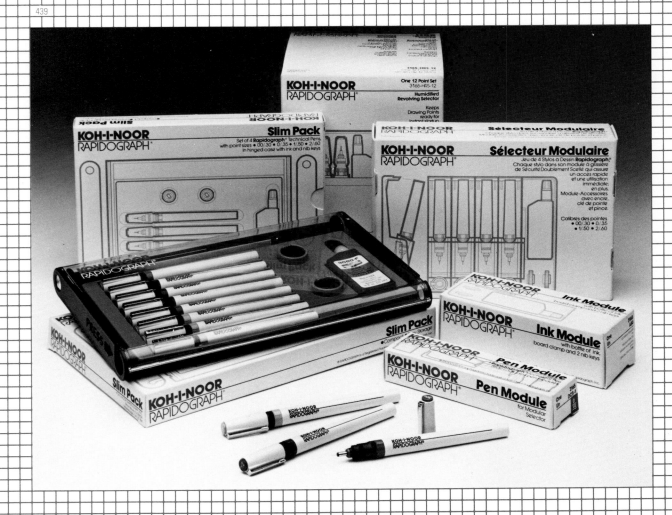

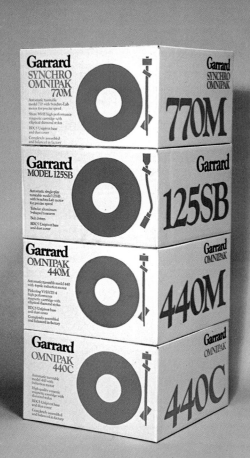

440. **Product:** Garrard Packers
Design Firm: Martin Jaffe Design Inc.
Client: Garrard Division of Plessey Consumer Products

441. **Product:** Empire 200E Stylus
Account Manager: Owen W. Coleman
Designer: Owen W. Coleman/John Rutig
Design Firm: Coleman, LiPuma & Maslow, Inc.
Client: Empire Scientific Company

442. **Product:** Polaroid Film Line
Design Firm: Polaroid Coporation
Client: Polaroid Corporation

443. **Product:** Vivitar Exposure Meters
Designer: Tets Yamashita
Design Firm: Harte Yamashita & Forest
Client: Vivitar

444. Product: Gift Shorts (Underpants for Ladies)
     Art Director: Kensaku Iwaki
     Designer: Studio YAP
     Client: Wacoal Co., Ltd.

445. Product: Totes Bag Line
     Design Director: Ray Perszyk
     Designer: Jim Gabel
     Design Firm: Cato, Yasumura, Behaeghel Inc. (U.S.)
     Client: Totes

446. Product: Timex Watch Package
     Design Director: Clive Chajet
     Design Firm: Chajet Design Group Inc.
     Client: Timex Inc.

447. Product: Enjoy Golf Panties
     Art Director: Kensaku Iwaki
     Designer: Studio YAP
     Client: Wacoal Co., Ltd.

448. Product: Jean Patou Chausettes
     Art Director: Masahiro Oishi
     Designer: Masahiro Oishi/Kyoichi Nakagawa
     Client: Unitika Berkshire Co., Ltd.

446

447

448

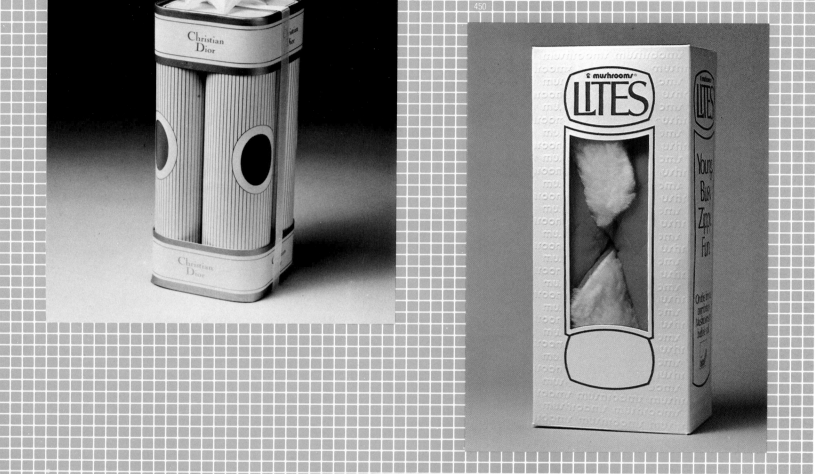

449. **Product:** Christian Dior Bas Collants
Art Director: Masao Sano
Designer: Osamu Nakabayashi
Client: Kanebo Dior Co., Ltd.

450. **Product:** Mushroom Lites Slippers
Art Director: Stewart Mosberg/James Beukelaer
Design Firm: Stewart Mosberg Design Associates, Inc.
Client: R.G. Barry Corp Mushrooms Division

451. **Product:** Saddlebred Shoe Box
Design Firm: Old Dominion Box Company, Inc.
Client: Craddock-Terry Shoe Company

452. **Product:** Ace Comb Line
Design Firm: Si Friedman Associates, Inc.
Client: Ace Comb Co.

453. **Product:** Munsingwear Men's Underwear
Art Director: Richard Gerstman
Designer: Vicki Cero
Design Firm: Gerstman & Meyers, Inc.
Client: Munsingwear, Inc.

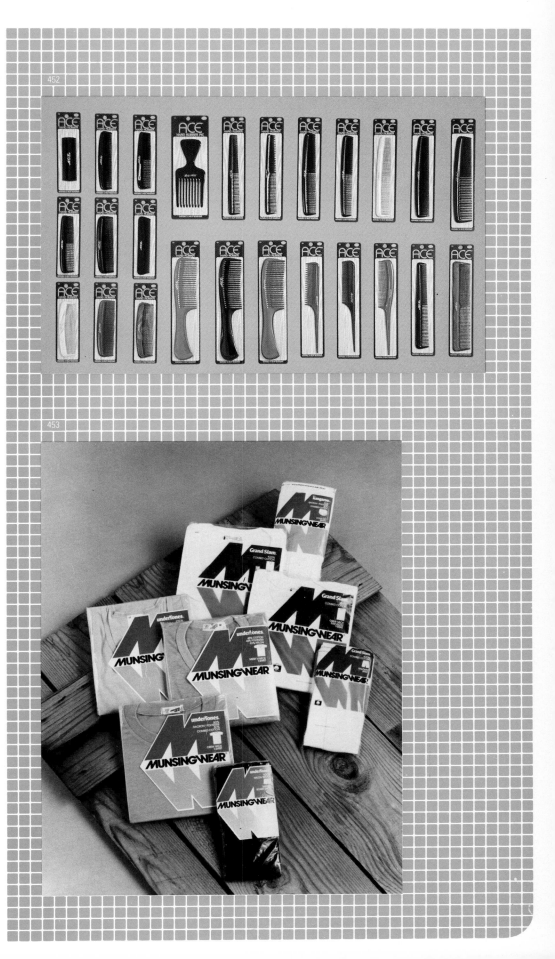

454. Product: Bausch & Lomb Ski Goggles (old)
     Client: Bausch & Lomb Inc.

455. Product: Bausch & Lomb Ski Goggles (redesign)
     Account Director: Owen W. Coleman
     Designer: Owen W. Coleman/John Rutig
     Design Firm: Coleman, LiPuma & Maslow, Inc.
     Client: Bausch & Lomb Inc.

456. **Product:** Weyless Bicycle Gear
Design Firm: Charles Biondo Design Associates, Inc.
Client: Weyless Industries, Inc.

457. **Product:** Let's Go Fishing Fishing Pole
Creative Director: Arthur Eilertson
Designer: Frank Weitzman
Design Firm: Gregory Fossella Associates
Client: Ryobi America

458

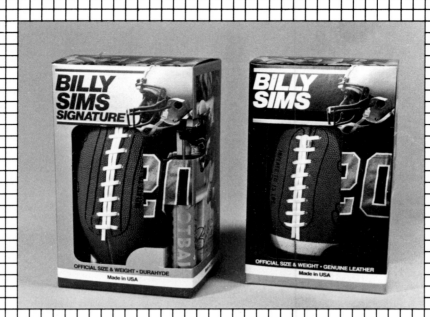

459

460

458. Product: Billy Sims Football
Designer (graphic): Arnold A. Martin Associates
Designer
(structural): Thomas J. Sellors, Olympic Packaging, Inc.
Design Firm: Olympic Packaging, Inc.
Client: Acme Sporting Goods

459. Product: Skinny Dip Clip
Creative Director: Martin Beck
Designer: George Boesel
Design Firm: Gregory Fossella Associates
Client: Skinny Dip Clip

460. Product: Ultracast Reel
Creative Director: Arthur Eilertson
Designer: Frank Weitzman
Design Firm: Gregory Fossella Associates
Client: Ryobi America

461. Product: Shopping Bags (Seryo)
Art Director: Shiro Tazumi
Designer: Shiro Tazumi
Client: Seryo Inc.

462. Product: Gallery Originals
Art Director: Timothy J. Musios
Designer: Neil T. Davis
Design Firm: Avon Products, Inc.
Client: Avon Products, Inc.

461

462

463. **Product:** Gilmore's Shopping Bags
Art Director: Ray Peterson
Designer: Marvin Steck
Design Firm: Container Corporation of America
Client: Gilmore's

464. **Product:** Toshiba Shopping Bag
Art Director: Koji Mizutani
Designer: Koji Mizutani
Client: Toshiba Co., Ltd.

465. **Product:** Avignon Freres Caterers Lunchbox
Designer: Sam Gass/Fred Mark
Design Firm: Lebanon Packaging Corporation
Client: Avignon Freres

466. **Product:** Gilmore's Dept. Store Boxes
Art Director: Ray Peterson
Designer: Marvin Steck
Design Firm: Container Corporation of America
Client: Gilmore's

467. **Product:** Kakiyama Shopping Bag
Art Director: Kazuaki Iwasaki
Designer: Kazuaki Iwasaki
Client: Shinsaibashi Kakiyama Inc.

465

466

467

468. Product: Royal Doulton Egg Box
Designer: Nancy E. Oliver-Clarke
Design Firm: The Chaspec Manufacturing Company
Client: Royal Doulton

469. Product: Hotel New Carina Wrapping Paper
Art Director: Takeo Yao
Design Firm: Yao Design Institute, Inc.
Client: Hotel New Carina

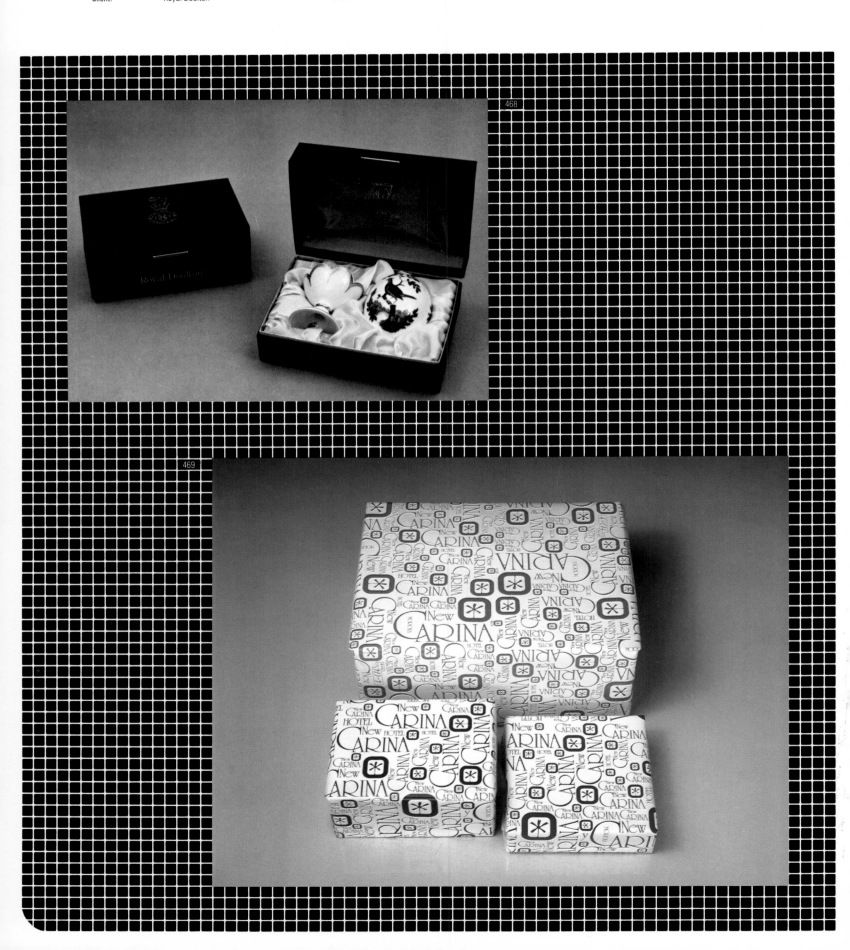

470. Product: Christmas 1982 Corporate Packaging
     Statement
     Art Director: Timothy J. Musios
     Designer: Neil T. Davis/Ronald W. Longsdorf
     Design Firm: Avon Products, Inc.
     Client: Avon Products, Inc.

471. Product: Norman Rockwell Plate Box
     Designer: Ed Gillies Marketing, Inc.
     Design Firm: The Chaspec Manufacturing Company
     Client: River Shore, Ltd.

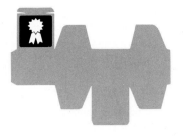

Award Winning Packaging

"One biscuit firm I used to work for sells in 140 markets, from Fiji to the Falkland Islands. It, like others, collects countries like stamps! Design is an indispensable investment for real long-term multi-marketing."

James Pilditch

Design and packaging groups in many nations sponsor annual contests to determine the best packages; the type of award as well as the breadth of the contest varies from nation to nation. By closely examining the award winners in each of these contests, it is possible to establish national preferences in packaging. Concentration tends to be on packaging for those products which are either exported from the nation or grown or produced there on an especially large scale.

Certain national design characteristics become evident. The Germans are definitely pioneers in clearing the clutter from graphic design. Multilingual nations such as Canada and South Africa have to cope with the additional problem of working with two languages. The British, Japanese and Swiss are beginning to produce some excellent package designs.

Throughout Europe there is a growing awareness of consumer needs. In the United Kingdom, there are more package design firms than anywhere outside the United States. The number of professional independent designers is also increasing in the Netherlands, Sweden and Denmark. On the other hand, it is interesting to note that Germany and France appear to favor designs submitted by package suppliers and advertising agencies.

National Color Preferences

Various nationalities have different color associations. The market served by the specific product is of extreme importance in selecting a suitable package color. Social strata, age, race and whether the target population is urban or rural are also important factors in determining color selection.

In the United States, red is indicative of cleanliness, while in the United Kingdom the same color is the least clean of all. In Sweden, blue is a masculine color, but in the Netherlands blue is feminine. In Italy, a land with plenty of sunshine, red is a popular color; the United Kingdom, Sweden and the Netherlands prefer blue and golden yellow. Blue connotes "seriousness" in the Netherlands, while in the United States green is most likely to have this effect. Red is a very popular choice for Balkan nations.

An interesting observation is that color preferences change in relation to the state of the economy. In prosperous times, red and blue, indicative of strength and vitality, sell best; in less booming economies, harmonious and serious colors such as yellow-green are preferred.

In an evaluation carried out by the Polish government, it was found that older people prefer to buy children's products in pink and blue packages, but that younger buyers prefer light green and warm yellow Thus, an item intended for a "grandparent" market should be packaged in different colors than one geared to parents or young people.

A basic rule of thumb is that in nations having bright sunlight, colors should be strong. Under the influence of tropical light, pastel shades tend to appear old and musty. This fact can and should be considered when using a color to convey a particular message.

Object Symbolism

Just as important as color is the symbolism of objects and the printed message. In Taiwan, the elephant denotes strength and would be a desirable symbol for a package. But in nearby Thailand, the elephant is the national emblem and its use on a package would be equivalent to use of the Stars and Stripes in the United States. In Islamic nations, pigs should be

avoided for religious reasons and six-pointed stars such as the Star of David for political reasons. Cows should be avoided in Hindu nations.

Certain symbols are simply meaningless in many areas. Snowflakes are meaningful to people who have seen snow, but connote little or nothing to inhabitants of tropical areas. The mushroom is a symbol of good luck in Central Europe, but meaningless elsewhere.

International Packaging and the United States

International products and their packages are of direct interest to the large importing nations such as the United States. Imported foods, in particular, are of great interest to Americans, who travel fairly widely and have sampled foreign cuisines. Also, with the trend toward

smaller families, increasing numbers of adults are willing to pursue sophisticated tastes in food as a cultural event, not merely as a biological necessity. A new American cuisine is emerging which is strongly influenced by both European and Oriental traditions. Both international packages and their products are major factors in shaping the future of this new cuisine.

PACKAGING DESIGN TRENDS IN EUROPE AND NORTH AMERICA

NATION	CURRENT TRENDS	IMPROVEMENTS NEEDED
Canada	Conservative, high level of taste. Less swayed by flamboyance. Busy-appearing packages because of dual language.	Better control of legislative demands on design.
Denmark	More advanced than U.S. design. Simple, straight design lines. Extensive use of corporate symbols.	Better artwork and printing on all materials.
France	Nostalgic look popular, unlike U.S. packaging. Not a leader in innovative design.	Simplified line and graphics to unify looks of product lines.
Germany	Imaginative use of new materials and systems. Extensive use of plastics. Little use of corporate symbols on packages.	Improved consumer convenience and communication.
Japan	Superb graphics. Excellent use of natural materials. Good, clear copy. Leader in Asian package design.	Less overpackaging; more automation.
Netherlands	Excellent surface graphics. Rapid introduction of new products and packages. Strong pan-European influence.	More convenience items. Simplification of product names.
Norway	Good graphics. Excellent designs on most packages.	Need to make packages more consumer-oriented.
Sweden	Superb package design. Modern trends followed.	Better nostalgic packaging.
United Kingdom	Growing design industry. Less garish, more discreet design than in U.S.	More important view of package in overall business cycle. Improved large-sized packages.
United States	Simple, clear graphics; uncluttered letters; bold design. Extensive use of natural, earthy colors. More use of photography on labels and packages.	Shelf-stable products, more convenience foods. Economical materials and more easily disposable packages.

472

473

474

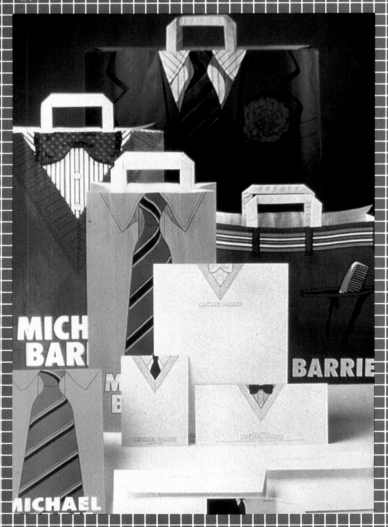

472. Product: Chaps Cologne
     Award: Clio '82
     Award Organization: Clio Awards
     Country: United States
     Designer: Primary Design Group
     Client: Warner Lauren

473. Product: Pickle's Gin
     Award: Clio '82
     Award Organization: Clio Awards
     Country: United States
     Designer: Mathieu, Gerfen & Bresner
     Client: Canadian Schenley Inc.

474. Product: Michael Barrie Bags
     Award: Clio '82 International
     Award Organization: Clio Awards
     Country: England
     Designer: Michael Peters, London
     Client: Michael Barrie Menswear

475. Product: Noallec Line
     Award: Clio '82 International
     Award Organization: Clio Awards
     Country: Japan
     Designer: Albion Cosmetics Co.
     Client: Albion Cosmetics Co.

476. Product: Quick Fix Line
     Award: Clio '82
     Award Organization: Clio Awards
     Country: United States
     Designer: Robert P. Gersin Associates, Inc.
     Client: General Electric Company

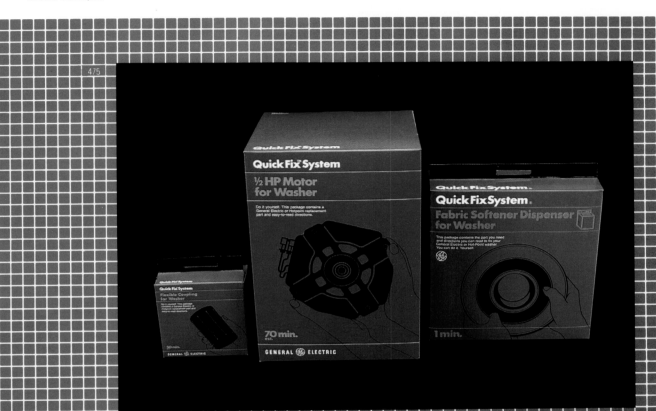

477

478

479

480

481

482

477. Product: Animal Crackers Kit
Award: Clio '82
Award Organization: Clio Awards
Country: United States
Designer: Robert P. Gersin Associates, Inc.
Client: Ginn & Company

478. Product: Animal Crackers Kit
Award: Clio '82
Award Organization: Clio Awards
Country: United States
Designer: Robert P. Gersin Associates, Inc.
Client: Ginn & Company

479. Product: Showa Gift Set
Award: Clio '82 International
Award Organization: Clio Awards
Country: Japan
Designer: Design Studio Onion Soup, Tokyo
Client: Showa

480. Product: Nutri Grain Cereal
Award: Clio '82
Award Organization: Clio Awards
Country: United States
Designer: J. Walter Thompson, Chicago
Client: Kellogg Company

481. Product: Kraft a la Carte Dinners
Award: Clio '82
Award Organization: Clio Awards
Country: United States
Designer: N.W. Ayer
Client: Kraft, Inc.

482. Product: The Goodman System
Award: Clio '82
Award Organization: Clio Awards
Country: United States
Designer: Bob Frissora Design Company
Client: The Goodman System Company

483

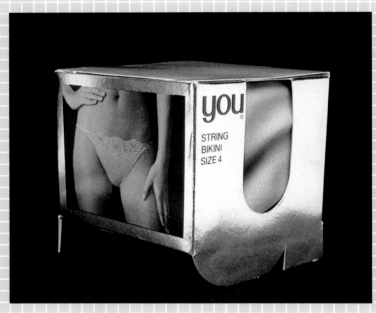

484

485

483. **Product:** You Panties
Award: 1981 PDC Gold
Award Organization: Package Designers Council
Country: United States
Designer: N.W. Ayer Design
Client: Formfit

484. **Product:** Designer Salami
Award: 1981 PDC Gold
Award Organization: Package Designers Council
Country: United States
Designer: Primo Angeli Graphics
Client: P.G. Molinari & Sons

485. **Product:** Herman Joseph's Beer
Award: 1981 PDC Gold
Award Organization: Package Designers Council
Country: United States
Designer: DeMartin-Marona-Cranstoun-Downes
Client: Adolph Coors Company

486. **Product:** Cutlery & Cutting Board Sets
Award: 1981 PDC Gold
Award Organization: Package Designers Council
Country: United States
Designer: J.C. Penney
Client: J.C. Penney

487. **Product:** Liquid Creme Soap
Award: 1981 PDC Gold
Award Organization: Package Designers Council
Country: United States
Designer: Robertz, Webb and Company
Client: Jovan, Inc.

486

487

488

489

490

491

488. **Product:** Noodles by Leonardo
Award: 1981 PDC Gold
Award Organization: Package Designers Council
Country: United States
Designer: Container Corporation of America
Client: Noodles by Leonardo

489. **Product:** Koh-I-Noor Rapidograph Pens
Award: 1981 PDC Gold
Award Organization: Package Designers Council
Country: United States
Designer: Robert Hain Associates, Inc.
Client: Koh-I-Noor Rapidograph

490. **Product:** Certicare Line
Award: 1981 PDC Gold
Award Organization: Package Designers Council
Country: United States
Designer: Selame Design
Client: Amoco Oil Co.

491. **Product:** Rich's Dessert Line
Award: 1981 PDC Gold
Award Organization: Package Designers Council
Country: United States
Designer: Landor Associates, NY
Client: Rich Product Corp.

492. **Product:** The Handle
Award: PPC Silver
Award Organization: Paperboard Packaging Council
Country: United States
Designer: Packaging Corporation of America
Client: Rocky Mountain Tools

493. **Product:** Dansk Bistro Collection
Award: PPC Silver
Award Organization: Paperboard Packaging Council
Country: United States
Designer: Rand-Whitney Packaging Corporation
Client: Dansk International Design, Ltd.

494

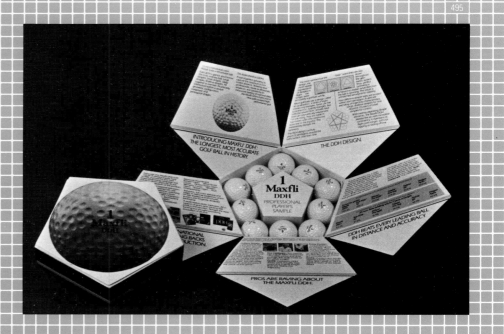

495

496

497

494. Product: Conture Condoms
     Award: PPC Silver
     Award Organization: Paperboard Packaging Council
     Country: United States
     Designer: Package Products Company
     Client: Akwell Industries

495. Product: Maxfli Golf Balls
     Award: PPC Silver
     Award Organization: Paperboard Packaging Council
     Country: United States
     Designer: Etta Packaging, Inc.
     Client: Dunlop Sports Co.

496. Product: Gypsy Moth Traps
     Award: PPC Gold
     Award Organization: Paperboard Packaging Council
     Country: United States
     Designer: Old Dominion Box Co., Inc.
     Client: Reuter Laboratories

497. Product: Avon '81 Christmas Box Line
     Award: PPC Silver
     Award Organization: Paperboard Packaging Council
     Country: United States
     Designer: Avon Products, Inc.
     Client: Avon Products, Inc.

498. Product: Functional Lighting
     Award: PPC Special Citation
     Award Organization: Paperboard Packaging Council
     Country: United States
     Designer: Berles Carton Company, Inc.
     Client: C.N. Burman Co., Heldak Lighting Products Corp.

499. Product: Privilege Chocolates
     Award: PPC Silver
     Award Organization: Paperboard Packaging Council
     Country: United States
     Designer: Field Sons & Co., Ltd. England
     Client: Tobler Suchard Limited

500. Product: Table Delights
     Award: PPC Silver
     Award Organization: Paperboard Packaging Council
     Country: United States
     Designer: Rand-Whitney Packaging Corp.
     Client: Colony

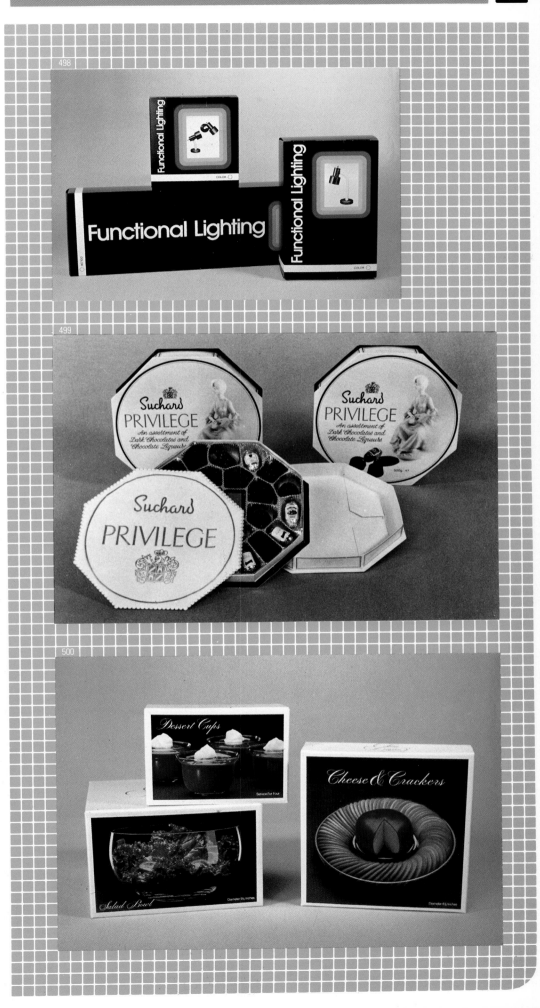

501. **Product:** Crisco Oil
Award: Winner '81-'82
Award Organization: Flexible Packaging Association
Country: United States
Designer: Champion International

502. **Product:** Sine-Off
Award: Winner '81-'82
Award Organization: Flexible Packaging Association
Country: United States
Designer: Reynolds Metals Co.

503. **Product:** Similac Powder
Award: Winner '81-'82
Award Organization: Flexible Packaging Association
Country: United States
Designer: Reynolds Metals Co.
Client: Ross

504. **Product:** McCain Country Style Wedges
Award: Winner '81-'82
Award Organization: Flexible Packaging Association
Country: United States
Designer: Crown-Zellerbach

505. **Product:** Bill Blass Eyewear Box
Award: Excellence Award: Eyewear
Award Organization: National Paperbox & Packaging Association
Country: United States
Designer: Shaw Packaging Inc.
Client: Universal Optical Co., Inc.

506. **Product:** SSAC Designer's Kit Box
Award: Excellence Award: Electronic & Photographic Products
Award Organization: National Paperbox & Packaging Association
Country: United States
Designer: J.F Friedel Paper Box Company
Client: S.S.A.C., Incorporated

507. **Product:** Career Opportunities Boxes
Award: First Award: Educational Material
Award Organization: National Paperbox & Packaging Association
Country: United States
Designer: The Chaspec Manufacturing Company
Client: Time Share Corporation

508. **Product:** Sony Classics Eyewear
Award: Excellence Award: Eyewear
Award Organization: National Paperbox & Packaging Association
Country: United States
Designer: F.H. Buffinton Company
Client: Universal Optical Co., Inc.

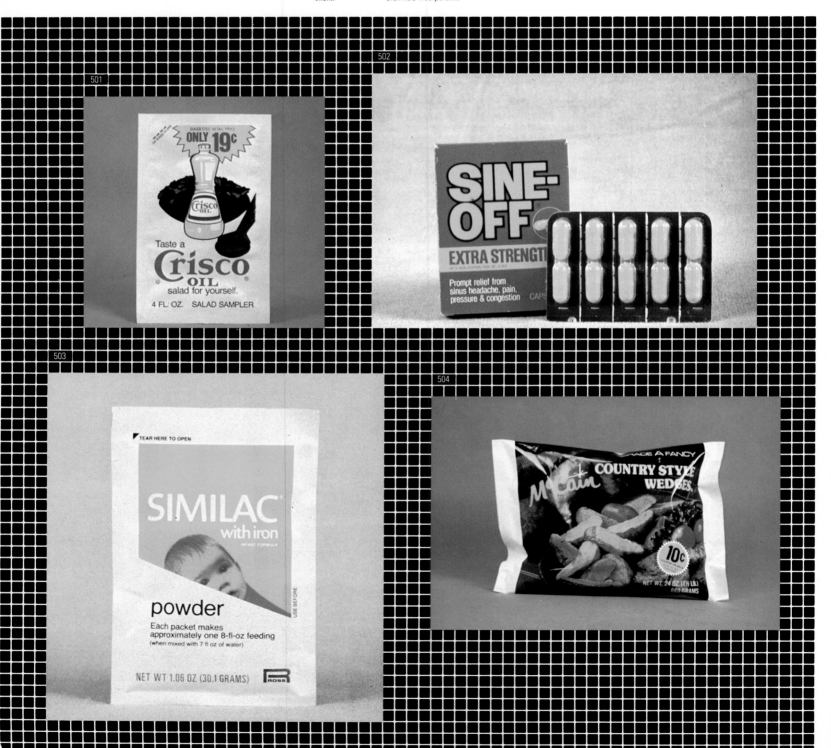

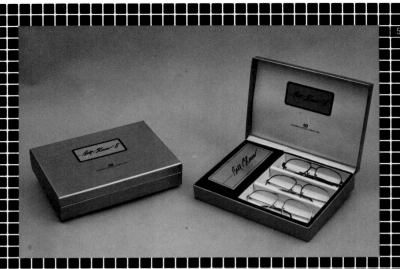

505

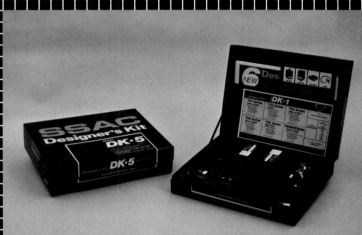

506

507

508

509. Product: Skandinavik Tobacco
     Award: AIMCAL '82
     Award Organization: Association of Industrial Metallizers,
     Coaters & Laminators
     Country: Denmark
     Designer: Raakmans Fabriker A/S Denmark
     Client: Skandinavisk Tobakskompagni

510. Product: Bachman Pretzels
     Award: AIMCAL '82
     Award Organization: Association of Industrial Metallizers,
     Coaters & Laminators
     Country: United States
     Designer: Bachman Co.
     Client: Bachman Co.

511. Product: Prosciutto Santa Rosa
     Award: AIMCAL '82
     Award Organization: Association of Industrial Metallizers,
     Coaters & Laminators
     Country: Canada
     Designer: Santa Rosa Food, Ltd.
     Client: Santa Rosa Food, Ltd.

512. Product: Blue Stratos Cologne for Men
     Award: AIMCAL '82
     Award Organization: Association of Industrial Metallizers,
     Coaters & Laminators
     Country: United States
     Designer: Shulton, Inc.
     Client: Shulton, Inc.

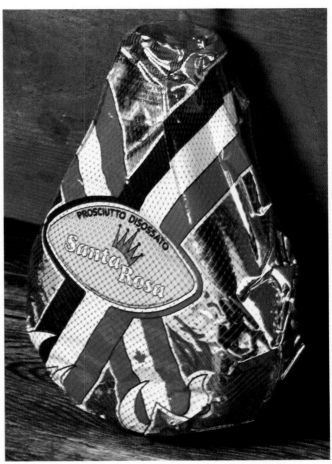

511

512

513. Product: Kiesenbosch Keg
     Award: IPSA TArophy
     Award Organization: The Institute of Packaging South Africa
     Country: South Africa
     Designer: Kohler Plastics
     Client: Kiesenbosch

514. Product: Buendia Coffee
     Award: Gold Medal
     Award Organization: Leipzig Spring Fair, March 1982
     Country: Colombia
     Designer: Mr. Jaime Gutierrez Lega
     Client: Cafe Leofilizado

515. Product: Krauterbox Seeds
     Award: '81 Austrian
     Award Organization: Austrian Packaging Institute
     Country: Austria
     Designer: E. Schausberger & Co.
     Client: Austrostaat

516. Product: Wiener Mode Bon-Bons
     Award: '81 Austrian
     Award Organization: Austrian Packaging Institute
     Country: Austria
     Designer: Lucie Bucheim
     Client: Wiener Modezeitschrift

513

514

515

516

517. **Product:** Fruit Alcohols
Award: Golden Chestnut-Certificate of Merit
Award Organization: Polish National Packaging Competition
Country: Poland
Designer: B. Paleta

518. **Product:** Avalanche Photo Diodes
Award: Golden Nut Award
Award Organization: Netherlands Packaging Institute
Country: Netherlands
Designer: Verpakkingsontwerpbureau, Veldhoven
Client: Werkplaats Verpaakingsontwerpbureau Elcoma

519. **Product:** Valflora Plant Package
Award: Golden Nut Award
Award Organization: Netherlands Packaging Institute
Country: Netherlands
Designer: Van Leer Plastics, Amsterdam
Client: Koninklijke Emballage Industrie Van Leer BV

520. **Product:** Network Cargo Systems
Award: Golden Nut
Award Organization: Netherlands Packaging Institute
Country: Netherlands
Designer: NV Philips' Gloeilampenfabrieken, Eindhoven
Client: Werkplaats Verpakkingsontwerpbureau Elcoma

517

518

519

520

521.	Product:	Rustico Powder
	Award:	Gold (Corrugated Fibreboard) '81
	Award Organization:	Packaging Association of Canada
	Country:	Canada
	Designer:	SPB Canada, Inc.
	Client:	Sico Inc.

522.	Product:	Oster Health Center
	Award:	Gold (Corrugated Fibreboard) '81
	Award Organization:	Packaging Association of Canada
	Country:	Canada
	Designer:	B.I. Graphics
	Client:	Oster Division of Sunbeam Corp. (Canada) Ltd.

523.	Product:	Carlsberg Gold Carton
	Award:	Gold (Corrugated Fibreboard) '81
	Award Organization:	Packaging Association of Canada
	Country:	Canada
	Designer:	Carling O'Keefe Breweries of Canada Limited
	Client:	Carling O'Keefe Breweries of Canada Limited

524.	Product:	King Kiwi Strawberries
	Award:	Winner of Consumer Packs Section
	Award Organization:	New Zealand Forest Products Awards
	Country:	New Zealand
	Designer:	UEB
	Client:	King Kiwi

525.	Product:	Bowater Containers
	Award:	Gold Star
	Award Organization:	The Institute of Packaging
	Country:	England
	Designer:	Bowater Containers
	Client:	Bowater Containers

526.	Product:	The Picnic Sandwich
	Award:	Gold Star (Student Starpack Design)
	Award Organization:	The Institute of Packaging
	Country:	England
	Designer:	Mr. Crispin Ellis

527.	Product:	Stomoxin/Games Cube
	Award:	Most Innovative Paper or Paperboard Package
	Award Organization:	New Zealand Forest Products Packaging Awards
	Country:	New Zealand
	Designer:	Cunningham Packaging Ltd.
	Client:	Stomoxin/Games Cube

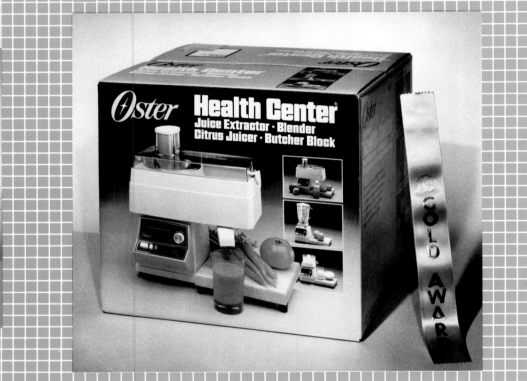

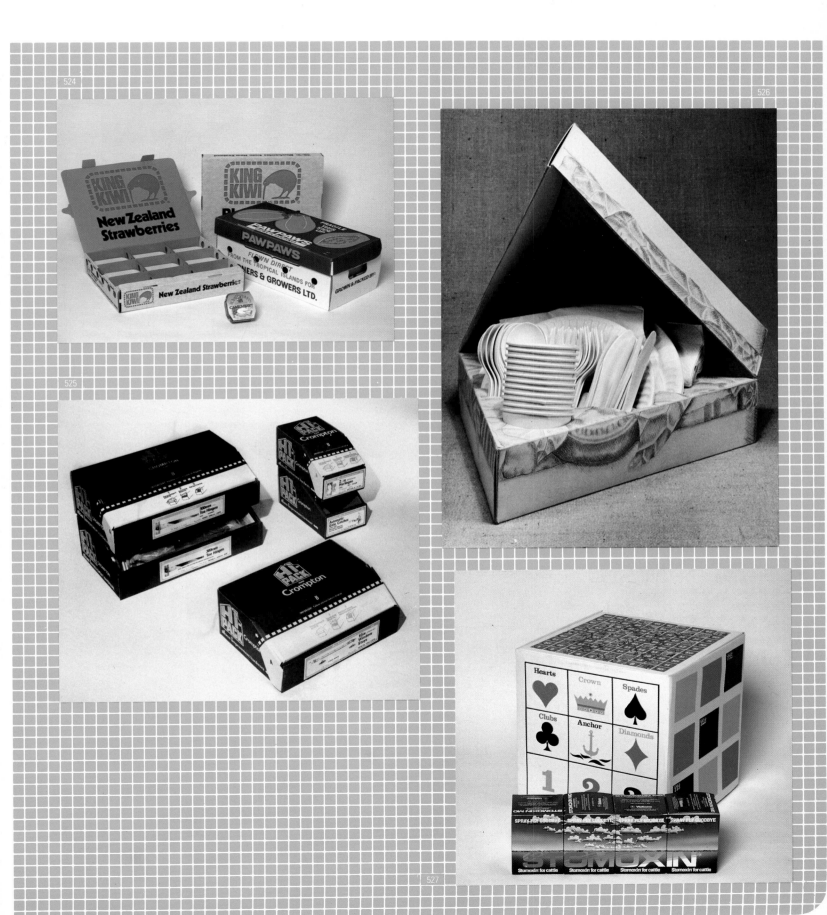

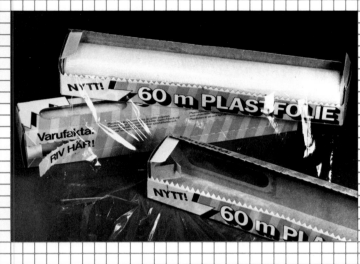

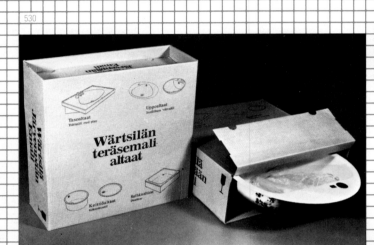

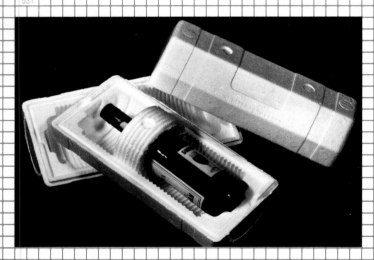

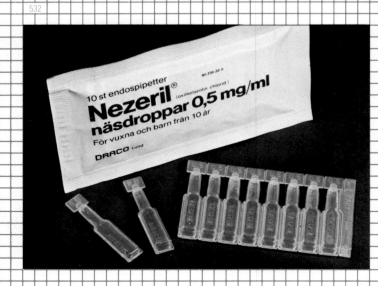

528. **Product:** Bookpack
Award: Scanstar '81
Award Organization: Scanstar
Country: Finland
Designer: Serlachius forpackningsgrupp
Client: Oy Valitut Palat- Reader's Digest Ab

529. **Product:** Plastic Film Dispenser
Award: Scanstar '81
Award Organization: Scanstar
Country: Sweden
Designer: Kartongfabriken Excelsior AB
Client: Ceson Plast AB

530. **Product:** Transport and Sales Pack
Award: Scanstar '81
Award Organization: Scanstar
Country: Finland
Designer: Oy Tampella AB
Client: Oy Wartsila Ab

531. **Product:** Bottle Transportation Pack
Award: Scanstar '81
Award Organization: Scanstar
Country: Denmark
Designer: Colon Colopor
Client: Colon Colopor

532. **Product:** Endospippet
Award: Scanstar '81
Award Organization: Scanstar
Country: Sweden
Designer: AB Astra
Client: AB Draco

533. **Product:** Post Pack for China
Award: Scanstar '81
Award Organization: Scanstar
Country: Norway
Designer: Norpapp Industri
Client: Porsgrund Porselaensfabrik

534. **Product:** Apple Box with Handle
Award: Scanstar '81
Award Organization: Scanstar
Country: Denmark
Designer: Colon vest
Client: Gartnernes Salgsforening

535. **Product:** Fish Case
Award: Scanstar '81
Award Organization: Scanstar
Country: Finland
Designer: Yhtyneet Paperitehtaat Oy Paperituote
Client: Fiskare, fiskodlare, fiskhandlare i Finland

536. **Product:** Norema Corrugated Box
Award: Scanstar '81
Award Organization: Scanstar
Country: Norway
Designer: Norpapp Industri
Client: Norema A/S

537. Product: Hero Tray
Award: Fri Kopenskap Award '81
Award Organization: Forlags Ab Ask
Country: Sweden
Designer: Mat for Alla Storkok AB
Client: Hero Conserves

538. Product: Classic Coffee
Award: Consumer Package of Year Fri Kopenskap Award '81
Award Organization: Forlags Ab Ask
Country: Sweden
Designer: Arvid Nordquist HAB
Client: Esselte Pac

539. Product: Box for Biscuits and Chocolates
Award: Certificate of Merit-/Creative Idea
Award Organization: Packstar
Country: Hong Kong
Designer: Vicky Lee Mi-Kum

540. Product: The LBH Line
Award: Package of the Year- Fri Kopenskap Award '81
Award Organization: Forlags Ab Ask
Country: Sweden
Designer: Kronborsten AB
Client: Kronborsten AB

541. Product: Chinese Paint Set
Award: Certificate of Merit for Creative Idea
Award Organization: Packstar
Country: Hong Kong
Designer: Miss Alice Li Man Yan

537

542. Product: Egg Package
     Award: Certificate of Merit for Creative Idea
     Award Organization: Packstar
     Country: Hong Kong
     Designer: Miss Choi Lai Fun

543. Product: Canned Vegetables
     Award: Certificate of Merit for Creative Idea
     Award Organization: Packstar
     Country: Hong Kong
     Designer: Miss Risa Yeung Suk Kuen

544. Product: Fluorescent Lantern
     Award: Certificate of Merit for Paper
     Award Organization: Packstar
     Country: Hong Kong
     Designer: Choi Kai Yan/Ideal Design & Productions

545. Product: Lune Yogurt
     Award: '81 Winner
     Award Organization: German Institute of Packaging
     Country: West Germany
     Designer: Schupbach AG
     Client: Lunebest Molkerei

546. Product: Cache Plaque
     Award: Oscar de l'emballage
     Award Organization: Institut Francais de L'emballage
     Country: France
     Designer: Beghin Say

547. Product: Renault Tire Cover
     Award: Oscar de l'emballge
     Award Organization: Institut Francais de L'Emballage
     Country: France
     Designer: Cartonnages Reine
     Client: Regie Renault

548. Product: Cote d'or Candy
     Award: '81 Winner
     Award Organization: German Institute of Packaging
     Country: West Germany
     Designer: Gustav Stabernack
     Client: Cote d'or Deutschland

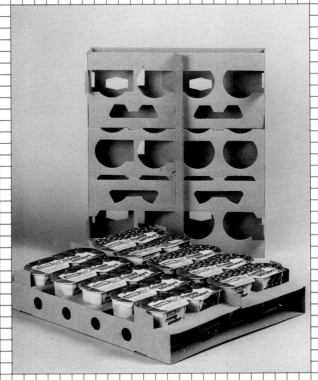

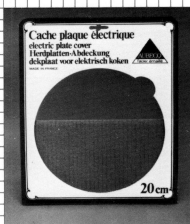

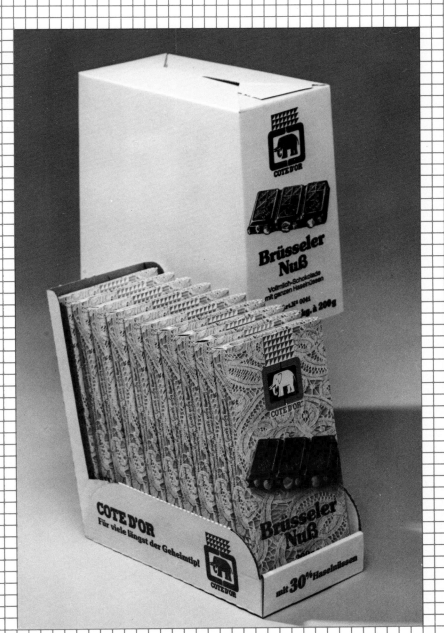

549. **Product:** Match Cosmetic Line
Award: '81 Winner
Award Organization: German Institute of Packaging
Country: West Germany
Designer: Therachemie
Client: Boes & Partner

550. **Product:** Penaten Puder
Award: Winner '81
Award Organization: German Institute of Packaging
Country: West Germany
Designer: Schmalbach-Lubeca
Client: Schmalbach-Lubeca

551. **Product:** Lunebest Dikmelk
Award: Winner '81
Award Organization: German Institute of Packaging
Country: West Germany
Designer: Schupbach
Client: Lunebest Molkerei

552. **Product:** Droste Chocolate Box
Award: Winner '81
Award Organization: German Institute of Packaging
Country: West Germany
Designer: Europa Carton
Client: Droste Fabrieken

553. **Product:** Boutique Cafe Coffee
Award: '81 Winner
Award Organization: German Institute of Packaging
Country: West Germany
Designer: Blechwarenfabriken Zuchner
Client: Verschiedene Kaffee-Grossrosterein

554. **Product:** Iglo Pekante Frozen Foods
Award: '81 Winner
Award Organization: German Institute of Packaging
Country: West Germany
Designer: Nicolaus Kempten
Client: Langnesse-Iglo

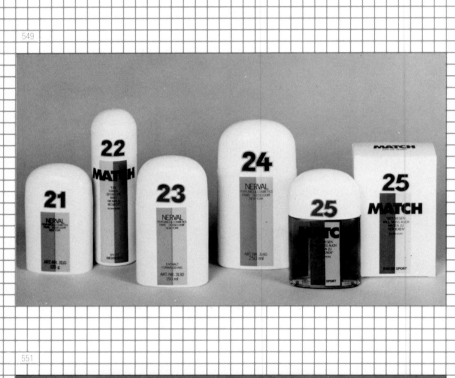

555. Product: Chocolate Drum Tin
     Award: Indiastar
     Award Organization: Indian Institute of Packaging
     Country: India
     Designer: Zenith Tin Works Private Limited

556. Product: Fruit Salad
     Award: '81 Winner
     Award Organization: German Institute of Packaging
     Country: West Germany
     Designer: Arbeitsgemeinschaft Gekalid
     Client: Nahrungsmittelindustrie

557. Product: Dispenser for Spices
Award: Indiastar
Award Organization: Indian Institute of Packaging
Country: India
Designer: Zenith Tin Works Private Ltd.

558. Product: Container for Spices
Award: Indiastar
Award Organization: Indian Institute of Packaging
Country: India
Designer: M/s. Brooke Bond India Ltd.
Client: M/s. Brooke Bond India Ltd.

559. Product: Cekatainer Pack for Vanaspati (a milk product)
Award: Indiastar
Award Organization: Indian Institute of Packaging
Country: India
Designer: Rollatainers Ltd.
Client: Gagan

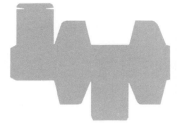

Afterword
Design In The Future

It wouldn't be too much a stretch of the imagination to anticipate a long-range trend to permanent secondary containers, elaborately decorated in whatever styles are appealing to the 21st century eye, the primary product unit being replaced as the product is consumed. Nor would it be too far-fetched to expect to find a wider range of reconstitutable food products of indefinite shelf life available to the consumer of the future than we ever dreamed possible.

It is in the area of manufacturing we are likely to see some of the greatest changes. Sharing the common concern with all machinery makers for the growing unreliability of energy sources, packaging equipment manufacturers will begin intense research over the next few decades for technical and geographic alternatives to improve the reliability of energy bases. Solar as well as matter-to-energy systems without the human dangers and pollution potential of atomic power should evolve. In addition, manufacturing and packaging concerns could concentrate on more favorable energy sources on distribution patterns.

In this context, it is not stretching the imagination too far to anticipate vast underground distribution networks, powered by pneumatic or other low-energy systems, as a means of transporting products to population centers. Major changes in vehicular types and uses seem a basic part of the alterations in the social fabric. These changes could be expressed not only in the vehicles now familiar to us, but also in the pioneering technology created to move us quickly and efficiently to the galactic systems we will surely populate in the decades beyond 2027.

From all of these elements, it is plain to see that the role of the designer, as well as all other individuals whose primary job is the conception, development, and the manufacture of packaging will be expanding dramatically.

Roy Parcels
Modern Packaging

Appendix
International Packaging Associations and Awards Organizations

Argentina

Instituto Argentino del Envasse
Avda. Belgrano 2852
1209—Buenos Aires

Australia

The Australian Institute of Packaging
P.O. Box 20
Chatswood, N.S.W.

Packaging Council of Australia
370 St. Kilda Road
Melbourne, Vic. 3004

Austria

Österreichisches Institut für Verpackungs Wesen
Geschaftselle: Gumpendorferstrasse 6
A-1060 Vienna

Österreichisches Verpackungs—Zentrum
Hoher Markt 3
A-1011 Vienna

Belgium

Institut Belge de l'Emballage
Rue Picard 15
1020 Brussels

Brazil

Associacão Brasileira de Embalagem
Av. Paulista, 688-15°
São Paulo 3

Canada

Packaging Association of Canada
10 St. Mary Street
Toronto, Ontario, M4Y 1P9

Chile

Instituto del Empaque de Chile
Pedro de Valdivia 1481
Santiago

Colombia

Department de Empaques y Embalajes
Proexpo, Apartado Aereo 19766
Bogota, D.E.

Cuba

Comisión Nacional de Envases y Embalajes
Calle 3re No. 3605
Esq. A, 36-A
Miramar, Havana

Czechoslovakia

Imados
U Micheleskeho Lesa 366
Prague

Democratic Republic of Germany

Zentralinstitut für Verpackungs—Wesen
Reisstrasse 42
8017 Dresden

Denmark

Emballageinstituttet
Jemtlandsgade 1
DK 2300 Copenhagen S

England

The Design Council
28 Haymarket
London SW1Y 45U

Flexible Packaging Association
31 Craven Street
London WC 2N5NP

The Institute of Packaging
Fountain House, 1A Elm Park
Stanmore, Middlesex HA 74BZ

Produce Packaging and Marketing Association
15 Hawley Square
Margate, Kent CT9 1PF

Egypt

**Egyptian Packaging Association
—PACKFORICO/EDPA**
P.O. Box 2408
Cairo

Federal Republic of Germany

**Institut für Lebensmitteltechnologie
und Verpackung**
Schragenhofstrasse 35
D-8000 Munich 50

Rationalisierungs—Gemeinschaft Verpackung
Postfach 11 91 93
6000 Frankfurt/Main

R G Verpackung im RKW
Gutleutstrasse 163-167, Postfach 11 91 93
6000 Frankfurt/Main

Finland

Finnish Packaging Association
Ritarikatu 3b A
SF-00170 Helsinki 17

France

Centre National de l'Emballage
Avenue Georges Politzer
78—Trappes

Institut Français de l'Emballage
40, rue du Colisée
75008 Paris

Hong Kong

Hong Kong Packaging Council
Eldex Industrial Building, 12/F, 21A Ma Tau
Hung Hom, Kowloon

Hungary

**Hungarian Institute of Materials Handling
and Packaging**
H-1431, P.O. Box 189
Budapest, Rigo u. 3

India

Indian Institute of Packaging
E-2 Marol Industrial Estate, Andheri East
Bombay 400093

Ireland

Irish Packaging Institute
Confederation House, Kildare Street
Dublin 2

Israel

The Israel Institute of Packaging
2 Carlebach Street, P.O. Box 20038
Tel Aviv, 61200

Italy

Instituto Italiano Imballaggio
Via Carlo Casan, 34
I-35100 Padova

Jamaica

Jamaica Packaging Association
8 Waterloo Road
Kingston 10

Japan

Japan Packaging Institute
Honshu Building, 12-8 5 Ginza Chuo-ku
Tokyo

Korea (South)

Korea Design and Packaging Centre
128 Yunkun-dong, Chongro-Ku; P.O. Box 23
Seoul

Mexico

Instituto Mexicano de Asistencía a la Industría
Homero 1425-602
Mexico 5, D.F.

Netherlands

Nederlands Verpakkingscentrum
P.O. Box 835
2501 CV, The Hague

New Zealand

New Zealand Institute of Packaging
Box 9130
Wellington
New Zealand

Norway

Den Norske Emballasjeforening
Klingenberg gt 7, Postbox 1754-Vika
Oslo 1

Pakistan

**The Packaging Cell of the Export
Promotion Bureau**
NPT Building
1. 1. Chundriger Road
Karachi Z

Peru

Instituto del Envase y Embalaje del Peru
Las Flores 346
San Isidro, Lima 27

Philippines

Asian Packaging Federation
Far East Building, Room 405, MCC P.O. Box 105
Makati, Rizal, 3117

Poland

Polish Packaging Research and Development Centre
ul. Konstancinska 11
02-942 Warsaw

Portugal

Centro Nacional de Embalagem
Praca das Industrias
Lisbon 3

Romania

Directia Pentru Ambalaje
Calea Victoriei nr 152
Sector 1, Bucharest

Spain

Instituto Español del Envase y Embalaje
Breton de los Herreros, 57
Madrid-3

Sri Lanka

Sri Lanka Institute of Packaging
c/o Aitken Spence Co., LTD
P.O. Box No. 5
Colombo

South Africa

The Institute of Packaging (S.A.)
P.O. Box 3259
Johannesburg 2000

Sweden

Svenska Forpackningsforsknings-institute
Box 91. Hammarby Fabriksvag 29-31
S-12122 Johanneshov 1

Swedish Packaging Institute
Box 9
S-16393 Spanga

Switzerland

Vereingung Schweizerches
Verpackungsinstitut
Verlang Max Binkert & Co.
CH-4335 Laufenburg

Vereinigung Schweiz
Verpackungsinstitut
Dreikonigstrasse 7
8002 Zurich

Taiwan

China Packaging Institute (C.P.I.)
489 Fu-Hsing N. Road
Taipei

Thailand

The Thai Packaging Association
Industrial Service Institute Building, Soi
Rama 4 Road, Bangkok 11

United States

Association of Industrial Metallizers, Coaters and Laminators (AIMCAL)
61 Blue Ridge Road
Wilton, CT 06897

Clio Awards
336 East 59th Street
New York, NY 10022

Container Corporation of America
400 East North Avenue
Carol Stream, IL 60187

Flexible Packaging Association
1090 Vermont Avenue, NW
Washington, D.C. 20005

Glass Packaging Institute
1800 K Street, NW
Washington, D.C. 20006

National Paperbox Association
231 Kings Highway
E. Haddonfield, NJ 08033

Package Designers Council
P.O. Box 3753
Grand Central Station
New York, NY 10017

Paperboard Packaging Council
Suite 600, 1800 K Street, NW
Washington, D.C. 20006

Union of Soviet Socialist Republics

All-Union Scientific Research, Experimental and Design Packaging Institute "WNIEKITU"
Gradcewskoje szose
Kaluga 9

Uruguay

Centro Uruguayo del Empaque
Sarandi 690 D-2° Entrepiso
Montevideo

Venezuela

Camara Venezolana del Envase
2° Piso-B, Esquina de Puente Anauco
Caracas

Yugoslavia

Packaging and Internal Transport Committee
Terazije 23
Belgrade